MATH STUDY SKILLS WORKBOOK

Second Edition

Your Guide to Reducing Test Anxiety and Improving Study Strategies

Paul D. Nolting, Ph.D.
Learning Specialist

Houghton Mifflin Co. • Boston New York

Publisher: Jack Shira
Senior Sponsoring Editor: Lynn Cox
Assistant Editor: Melissa Parkin
Associate Project Editor: Kristin Penta
Manufacturing Manager: Florence Cadran
Senior Marketing Manager: Ben Rivera
Marketing Assistant: Lisa Lawler

Cover art "The quilt of Gee's Bend": "Strips" circa 1875 by Anne May Young
© 2003 Tinwood Gallery

Printed in the U.S.A.

ISBN-13: 978-0-618-47303-8   ISBN-10: 0-618-47303-3
  7 8 9 10–B–09 08 07 06

Contents

Preface

Wouldn't it be nice if all we had to do was listen to a lecture on math and read the textbook in order to learn it? That would be paradise. However, most math courses take place on Earth, and students have to do much more than just take notes and read a textbook. They need a system of study skills that help them understand and master mathematics.

Many students in all levels of math courses have difficulty learning math because it is one of the most difficult subjects in college. First, many students who have struggled with math before going to college continue to struggle when they take their first college math courses, even developmental math courses. Second, some students who have done well in the past begin to struggle when they take upper level math like college algebra or calculus. They made A's and B's in previous math courses, but all of a sudden they are failing upper level math courses. These students probably lived off their intelligence until they took a math course that challenged them. Since they never had to study for math, they did not know where to begin to study. Students in all levels of math courses benefit from designing a system of study skills.

Does this sound far fetched? Not really. I have watched calculus students read this workbook and improve their grades in order to be more competitive candidates for engineering programs. For example, I asked two calculus II students why they were taking my math study skills course. I thought they needed a one credit hour course to allow them to graduate, but I was wrong. Both students made a C in calculus I and said that they needed to make an A or B in calculus II because they were engineering majors. In most cases, students who make C's in calculus are not admitted to engineering schools. Both students made an A in my course and a B in their calculus II course and went on to engineering school. Success for them!

Another success story belongs to a student who failed a beginning algebra course three times. Yes, three times. She came to my office ready to quit college all together. I convinced her to focus on designing a system of study skills before taking the course one more time. She did. With her new system of study skills, she passed with a B and stayed in college. Success for her!

Earning better math grades doesn't have to be the only benefit from spending time developing a system for studying math. According to many students, they were able to take the math study skills and adapt them to their other courses. In fact, some students claim it is easier than adapting general study skills to learning math. They reported that their other course grades also went up. What a good experience!

So, what kind of assistance do most students want? Students want tips and procedures they can use to help them improve their math grades. The math study suggestions, however, have to be based on research and be statistically proven to improve student learning and grades. **Math Study Skills Workbook** is based on **Winning at Math: Your Guide to Learning Mathematics Through Successful Study Skills** (2002), which is the only math study skills text that can boast statistical evidence demonstrating an improvement in students' ability to learn math and make better grades. Learning and using these study skills techniques will improve your grades.

Math Study Skills Workbook is designed to supplement your math course or study skills course or to be used as part of a math anxiety workshop. The workbook is designed for independent work or to be supplemented by lectures. To gain the most benefit, the workbook needs to be completed by midterm.

This workbook is designed to enhance learning by teaching math learning skills in small "chunks." After each section, you are required to recall the most important information by writing it down. The writing exercises are especially designed for you to personalize learning techniques, such as the ten steps to better test taking. Each chapter review is designed to reinforce your learning and to give you the opportunity to immediately select and use the best learning strategy from that chapter. Lecture Notes pages have been added, where possible, to allow you to include your notes in this workbook.

As you adapt and practice each study skill, place them into a larger system for studying math. Decide which study skills will help with each of the following tasks you must do when learning math:

1. Taking and reworking lecture notes
2. Learning vocabulary
3. Completing homework in a way that helps you learn and master the math
4. Mastering and memorizing the concepts
5. Preparing for tests
6. Taking tests
7. Managing any anxiety whether math or text anxiety

Remember, to reap the most benefit, you need to complete this workbook and be using your system of study skills by midterm. Practice and master! Then, maybe we can use your success story in our next edition of the workbook!

1

What You Need to Know to Study Math

Mathematics courses are not like other courses. Because they are different, they require different study procedures. Passing most of your other courses requires only that you read, understand, and recall the subject material. To pass math, however, an extra step is required: You must use the information you have learned to correctly solve math problems.

Learning general study skills can help you pass most of your courses, but special math study skills are needed to help you learn more and get better grades in math. In this chapter, you will find out

- why learning math is different from learning other subjects,
- what the differences are between high school and college math, and
- why your first math test is very important.

Why learning math is different from learning other subjects

In a math course, you must be able to do four things:

1. *Understand* the material
2. *Process* the material
3. *Apply* what you have learned to correctly solve a problem, and
4. *Remember* what you have learned to learn new material.

Of these four tasks, applying what you have learned to correctly solve a problem is the hardest.

Examples: Political science courses require that you learn about politics and public service. But your instructor will not make you run for

continued

governor to pass the course. Psychology courses require you to understand the concepts of different psychology theories. But you will not have to help a patient overcome depression to pass the course. In math, however, you must be able to correctly solve problems to pass the course.

Sequential Learning Pattern

Another reason learning math is different than learning other subjects is that it follows a *sequential learning pattern*, which simply means that the material learned on one day is used the next day and the next day, and so forth. This building-block approach to learning math is the reason it is difficult to catch up when you fall behind. *All* building blocks must be included to be successful in learning math.

You can compare learning math to building a house. Like a house, which must be built foundation first, walls second, and roof last, math must be learned in a specific order. Just as you cannot build a house roof first, you cannot learn to solve complex problems without first learning to solve simple ones.

Example: In a history class, if you study Chapters 1 and 2, do not understand Chapter 3, and end up studying and having a test on Chapter 4, you *could* pass. Understanding Chapter 4 in history is not totally based

continued

on comprehending Chapters 1, 2, and 3. To succeed in math, however, each previous chapter has to be completely understood before you can continue to the next chapter.

Sequential learning affects your ability to study for math tests, as well. If you study Chapter 1 and understand it, study Chapter 2 and understand it, and study Chapter 3 and *do not* understand it, then when you study for a test on Chapter 4, you are not going to understand it either, and you probably will not do well on the test.

REMEMBER

To learn the new math material for the test on Chapter 5, you must first go back and learn the material in Chapter 4. This means you will have to go back and learn Chapter 4 while learning Chapter 5. (The best of us can fall behind under these circumstances.) However, if you do not understand the material in Chapter 4, you will not understand the material in Chapter 5 either, and you will fail the test on Chapter 5. This is why the sequential learning of math concepts if so important.

The sequential learning pattern is also affected by

- your previous math course grade,
- your math placement test scores, and
- the time elapsed since your last math course.

Sequential learning is influenced by how much math knowledge you have at the beginning of your course. Students who forgot or never acquired the necessary skills from their previous math course will have difficulty with their current math course. If you do not remember what you learned in your last math course, you will have to relearn the math concepts from the previous course as well as the new material for the current course. In most other courses, such as the humanities, previous course knowledge is not required. However, in math you must remember what the last course taught you so that you are prepared for the current course. Measuring previous course knowledge will be explained in Chapter 2, "How to Discover Your Math-Learning Strengths and Weaknesses."

Math placement scores also affect sequential learning. If you barely scored high enough to be placed into a math course, then you will have math-learning gaps. Learning problems will occur when new math material is based on one of your learning gaps. The age of the placement test score also affects sequential learning. Placement test scores are designed to measure your *current* math knowledge and are to be used immediately.

Sequential learning is interrupted if math courses are taken irregularly. Math courses are designed to be taken one after another. By taking math courses each semester, without semester breaks in between courses, you are less likely to forget the concepts required for the next course. Research has shown that the more time between math courses, the more likely that a student will fail the math course.

Now that you understand the "building block" nature of math, think about your math history.

- What are your previous math grades?
- How well did you do on the math placement test at your college?

- How long has it been since you took a math course?
- When you look at your math history, are there semesters when you did not take math?

These questions are important because if there was too much time in between your different math courses, you may have forgotten important math concepts that you need in your current class. To use the building block analogy, the blocks may not be as strong any more.

Now that you understand how learning math is a building experience, what should you do? *First,* don't get anxious. Stay calm. *Second,* if your college has a diagnostic math inventory in the tutoring center or math lab, take it to see what math concepts you have forgotten. Then ask your instructor where you can go to relearn these math concepts. *Third,* take the time to follow through. Many students just give up too easily, or they think they will catch up a little at a time. Don't think that and don't give up. The energy put into your class at the beginning of the semester will be more productive than energy put into class at the end of the semester when you try to learn everything during the last week before the final exam. *Fourth,* study and really learn the math; don't practice mimicking it. *Finally,* when it is time to register for the next semester, register immediately so you will be able to get into the math class you need. Why do all this? Because math is sequential!

Math as a Foreign Language

Another helpful technique for studying math is to consider it a foreign language. Looking at math as a foreign language can improve your study procedures. If you do not practice a foreign language, what happens? You forget it. If you do not practice math, what happens? You are likely to forget it, too.

Students who excel in a foreign language must practice it *at least* every other day. The same study skills apply to math, because it is considered a foreign language.

Like a foreign language, math has unfamiliar vocabulary words or terms that must be put in sentences called *equations*. Understanding and solving a math equation is similar to speaking and understanding a sentence in a foreign language.

> **Example:** Math sentences use symbols (which stand for spoken words), such as
>
> = (for which you *say*, "equal"),
> - (for which you *say*, "less"), and
> *a* (for which you *say*, "unknown").

Learning *how* to speak math as a language is the key to math success. Currently, most universities consider computer and statistics (a form of math) courses as foreign languages. Some universities have now gone so far as to actually classify math as a foreign language.

Math as a Skill Subject

Math is a *skill subject*, which means you have to practice actively the skills involved to master it. Learning math is similar to learning to play a sport, learning to play a musical instrument, or learning auto mechanics skills. You can listen and watch your coach or instructor all day, but unless you *practice* those skills yourself, you will not learn.

> **Examples:** In basketball, the way to improve your free throw is to *see and understand* the correct shooting form and then *practice* the shots yourself. Practicing the shots improves your free-throwing percentage. However, if you simply listen to your coach describe the correct form and see him demonstrate it, but you do not practice the correct form yourself, you will not increase your shooting percentage.
>
> Suppose you want to play the piano, and you hired the best available piano instructor. You sit next to your instructor on the piano bench and watch the instructor demonstrate beginning piano-playing techniques. You see and understand how to place your hand on the keys and play. But what does it take to learn to play the piano? You have to place your hands on the keys and *practice*.
>
> Math works the same way. You can go to class, listen to your instructor, watch the instructor demonstrate skills, and understand everything that is said (and feel that you are quite capable of solving the problems). However, if you leave the class *and do not practice* — by working and successfully solving the problems — you will not learn math.

Many of your other courses can be learned by methods other than practicing. In social studies, for example, information can be learned by listening to your instructor, taking good notes, and participating in class discussions. Many students mistakenly believe that math can be learned the same way.

REMEMBER

Math is different. If you want to learn math, you must practice. Practice not only means doing your homework but also means spending the time it takes to understand the reasons for each step in each problem.

Math as a Speed Subject

Math is a *speed subject,* which means, in most cases, it is taught faster than your other subjects. Math instructors have a certain amount of material that must be covered each semester. They have to finish certain chapters because the next math course is based on the information taught in their courses. In many cases a common math department final exam is given to make sure you know the material for the next course. Instructors are under pressure to make sure you are ready for the final exam because it demonstrates how prepared you are for the next level in math. This is different from, let's say, a sociology course where if an instructor doesn't teach the last chapter it will not cause students too many problems in the next sociology or social science course. So don't complain to the math instructor about the speed of the course. Instead, improve your study skills so you can keep up!

Another way math is a speed subject is that most of the tests are timed, and many students think

that they will run out of time. THIS CAUSES PANIC AND FEAR! This is different than most of your other courses where you generally have enough time to complete your tests or in your other courses that have multiple choice tests where you can start bubbling the responses on the scantron sheet if you start running out of time. Students must not only understand how to do the math problems but also must learn the math well enough to complete the problems with enough speed to finish the test.

What makes me curious is, if students feel like they don't have enough time to complete the math test, why are most of them gone before the test time is over? Sure, students who have learned the math thoroughly may complete the test early. That makes sense. Some students leave, however, because they either don't know the material, want to leave the anxious environment, or carelessly work through the test.

So, since speed is an issue in learning math, what should you do? *First*, to use an analogy, start a daily workout program to stay in shape. Review, review, and review as you learn new material. *Second*, practice doing problems within a time constraint. Give yourself practice tests.

A Bad Math "Attitude"

Students' attitudes about learning math are different from their attitudes about learning their other subjects. Many students who have had bad experiences with math do not like math and have a bad attitude about learning it. In fact, some students actually *hate* math, even though these same students have positive experiences with their other subjects and look forward to going to those classes.

Society, as a whole, reinforces students' negative attitudes about math. It has become socially acceptable not to do well in math. This negative attitude has become evident even in popular comic strips, such as *Peanuts*. The underlying message is that math should be feared and hated and that it is all right not to learn math.

This "popular" attitude toward math may reinforce your belief that it is all right to fail math. Such a belief is constantly being reinforced by others.

The bad math attitude is not a major problem, however. Many students who hate math pass it anyway, just as many students who hate history still pass it. The major problem concerning the bad math attitude is how you use this attitude. If a bad math attitude leads to poor class attendance, poor concentration, and poor study skills, then you have a bad math attitude *problem*.

REMEMBER

Passing math is your goal, regardless of your attitude.

Section Review

1. How does a sequential learning pattern affect math learning?

_____

_____

_____

_____

_____

Give two examples:

Example 1: _____

_____

_____

_____

_____

_____

Example 2: _____

_____

_____

_____

_____

_____

2. List two examples of how learning math is similar to learning a foreign language.

Example 1: _____

_____

_____

_____

_____

Example 2: _____

_____

_____

_____

_____

3. How is math similar to a skills subject?

_____

_____

_____

_____

_____

4. List three ways a bad math attitude could affect your mathematics learning.

First Way: _____

_____

_____

Second Way: _____

_____

_____

Third Way: _____

_____

_____

5. Why is math study skills important at all levels of math (see Preface)?

_____

_____

_____

_____

_____

_____

6. Math study skills can help you in other subjects but general study skills usually cannot improve math learning. Why is this statement important? (see Preface)

_____

_____

_____

_____

_____

_____

The differences between high school and college math

Math, as a college-level course, is almost two to three times as difficult as high school-level math courses. There are many reasons for the increased difficulty: course class time allowance, the amount of material covered in a course, the length of a course, and the college grading system.

The first important difference between high school and college math courses is the length of time devoted to instruction each week. Most college math instruction, for the fall and spring semesters, has been cut to three hours per week; high school math instruction is provided five hours per week. Additionally, college courses cover twice the material in the same time frame as do high school courses. What is learned in one year of high school math is learned in one semester (four months) of college math.

Simply put, in college math courses you are receiving less instructional time per week and covering twice the ground per course as you were in high school math courses. The responsibility for learning in college is the student's. As a result, most of your learning (and *practicing*) will have to occur outside of the college classroom.

College Summer Semester Versus Fall or Spring Semester and the Difference Between Night and Day

College math courses taught during summer semesters are more difficult than those taught during fall or spring. Further, math taught during night courses is more difficult than math taught during day courses.

Students attending a six-week summer math session must learn the information — and master the skills — two and a half times as fast as students attending regular, full-semester math sessions. Though you receive the same amount of instructional classroom time, there is less time to understand and *practice the skills* between class sessions.

Summer classes are usually two hours per day, four days per week (nighttime summer classes are four hours per night, two nights per week).

Example: If you do not understand the lecture on Monday, then you have only Monday night to learn the material before progressing to more difficult material on Tuesday. During a night course, you have to learn and understand the material before the break; after the break, you will move on to the more difficult material — *that night.*

Because math is a sequential learning experience, where every building block must be understood before proceeding to the next block, you can quickly fall behind, and you may never catch up. In fact, some students become lost during the first half of a math lecture and never understand the rest of the lecture (this can happen during just one session of night class). This is called "kamikaze" math because most students do not survive summer courses.

If you *must* take a summer math course, take a 10- or 12-week *daytime* session so that you will have more time to process the material between classes.

Course Grading System

The grading system for math is different in college than in high school.

Example: While in high school, if you make a D or borderline D/F, the teacher more than likely will give you

continued

a D, and you may continue to the next course. However, in some college math courses, a D is not considered a passing grade, or if a D is made, the course will not count toward graduation.

College instructors are more likely to give the grade of N (no grade), W (withdrawal from class), or F for barely knowing the material. This is because the instructor knows that you will be unable to pass the next course if you barely know the current one.

Most colleges require students to pass two college-level algebra courses to graduate. In most high schools, you may graduate by passing one to three math courses. In some college-degree programs, you may even have to take four math courses and make a C in all of them to graduate.

The grading systems for math courses are very precise compared to the grading systems for humanities courses.

Example: In a math course, if you have a 79 percent average and you need 80 percent to get a B, you will get a C in the course. But if you make a 79 percent in English class, you may be able to talk to your instructor and do extra credit work to earn a B.

Because math is an exact science and not as subjective as English, do not expect your math instructor to let you do extra work to earn a better grade. In college, there usually is not a grade given for "daily work," as is often offered in high school.

In fact, *your test scores may be the only grades that will count toward your final grade.* Therefore, you should not assume that you will be able to "make up" for a bad test score.

The Ordering of College Math Courses

College math courses should be taken, *in order,* from the fall semester to the spring semester. If at all possible, avoid taking math courses from the spring to fall semesters. There is less time between the fall and spring semester for you to forget the information. During the summer break, you are more likely to forget important concepts required for the next course and therefore experience greater difficulty.

Section Review

1. Compare the amount of college class time to high school class time.

_____

_____

_____

_____

2. How do college summer math courses differ from courses offered in the spring or fall?

_____

_____

_____

_____

_____

3. How can the order of taking math courses affect your learning?

_____

_____

_____

_____

_____

_____

Why your first math test is very important

Making a high grade on the first major math test is more important than making a high grade on the first major test in other subjects. The first major math test is the easiest and, most often, the one the student is least prepared for.

Students often feel that the first major math test is mainly a review and that they can make a B or C without much study. These students are overlooking an excellent opportunity to make an A on the easiest major math test of the semester. (Do not forget that this test counts the same as the more difficult remaining math tests.)

At the end of the semester, these students sometimes do not pass the math course or do not make an A because of their first major test grade. In other words, the first math test score was not high enough to "pull up" a low test score on one of the remaining major tests.

Studying hard for the first major math test and obtaining an A offers you several advantages:

- A high score on the first test can compensate for a low score on a more difficult fourth or fifth math test. All major tests have equal value in the final grade calculations.

- A high score on the first test can provide assurance that you have learned the basic math skills required to pass the course. This means you will not have to spend time relearning the misunderstood material covered on the first major test while learning new material for the next test.

- A high score on the first test can motivate you to do well. Improved motivation can cause you to increase your math study time, which will allow you to master the material.

- A high score on the first test can improve your confidence for higher test scores. With more confidence, you are more likely to work harder on the difficult math homework assignments, which will increase your chances of doing well in the course.

What happens if after all your studying, you make a low score on your first math test? You can still use this test experience to help you improve your next grade or to help determine if you are in the right math course. Your first math test, no matter what you make on it, can be used as a diagnostic test. Your teacher can review your test with you to see which type of math problems you got right and which ones you need to learn how to solve. It may be that you missed only a few concepts that caused the low score, and you can learn how to do these problems by getting help from the teacher, a learning resource center, or the math lab. However, you need to learn how to do these problems immediately so that you don't fall behind in the course. After meeting with your teacher, ask for the best way you can learn the concepts that are the bases of the missed problems and how to prepare for the next test. Even students who made a B on the first math test can benefit by seeing the teacher.

If you made below a 50 on your first math test, I suggest that it might be a good idea to drop back to a lower level math course. Even though it might be beyond the first week of drop and add, many colleges/universities will let you drop back to a lower math class after the first major math test. Students who drop back get a good foundation in mathematics that helps them in their next math courses. On the other hand, I have seen students who insisted on staying in the math course and repeated it several times before passing. Some of the students stopped taking the math course and dropped out of college. Dropping back to a lower level math course and passing it is the smartest move. These students went on to become more successful in their math courses.

REMEMBER

It does not matter where you start as long as you graduate. Discuss this option with your teacher.

Section Review

1. Why is your first math test supposed to be the easiest?

   _____

   _____

   _____

   _____

   _____

   _____

2. Give three reasons why your first math test is so important.

   First Reason: _____

   _____

   _____

   _____

   _____

   Second Reason: _____

   _____

   _____

   _____

   _____

   _____

continued

Third Reason: _____

_____

_____

_____

_____

_____

3. What do you do if you fail your first test?

Option One: _____

_____

_____

_____

_____

_____

Option Two: _____

_____

_____

_____

_____

_____

Option Three: _____

_____

_____

_____

_____

_____

Chapter 1 Review

1. In a math course, you must be able to do three things. You must _____ the material, _____ the material, and _____ what you have learned to solve the problem.

2. Math requires _____ _____, which means one concept builds on the next concept.

3. Placement test scores are designed to measure your _____ math knowledge and are to be used _____.

4. Like a foreign language, math has unfamiliar vocabulary words or terms that must be put in sentences called _____.

5. Learning how to speak math as a _____ _____ is one key to math success.

6. Keeping a _____ attitude about math will help you study more efficiently.

7. College math courses are _____ to _____ times as difficult as high school math courses.

8. Math course are even more difficult than other courses because a grade of _____ or better is usually required to take the next course.

9. The math grading is exact; in many cases, you cannot do _____. _____ to improve your grade.

10. If you fail your first math test, you need to make an appointment with your _____ to review your _____. Then you can decide if you should _____ to a lower math course.

What is the most important information you learned from this chapter?

_____

_____

_____

_____

_____

_____

_____

How can you immediately use it?

_____

_____

_____

_____

_____

_____

_____

_____

Lecture Notes

How to Discover Your
Math-Learning Strengths and Weaknesses

Math-learning strengths and weaknesses affect students' grades. You need to understand your strengths and weaknesses to improve your math-learning skills.

Just as a mechanic does a diagnostic test on a car before repairing it, you need to do your own testing to learn what you need to improve upon. You do not want the mechanic to charge you for something that does not need repairing, nor do you want to work on learning areas that do not need improvement. You want to identify learning areas you need to improve or better understand.

In this chapter, you will learn

• how what you know about math affects your grades,
• how quality of math instruction affects your grades,
• how affective student characteristics affect your math grades,
• how to determine your learning style,
• how to develop a math-learning profile of your strengths and weaknesses, and
• how to improve your math knowledge.

Areas of math strengths and weaknesses include math knowledge, level of test anxiety, study skills, study attitudes, motivation, and test-taking skills. Before we start identifying your math strengths

and weaknesses, you need to understand what contributes to math academic success.

Dr. Benjamin Bloom, a famous researcher in the field of educational learning, discovered that your IQ (intelligence) and your cognitive entry skills account for 50 percent of your course grade.

Figure 1
Variables Contributing to
Student Academic Achievement

Cognitive Entry Skills
(how much math you know before entering a new math course)
and
IQ
(how fast you can learn old and new math concepts)
50%

Quality of Instruction
(Effectiveness of math instruction: textbook, teaching style, extra teaching aids, etc.)
25%

Affective Characteristics
(Personality, self-concept, locus of control, motivation, attitudes, anxiety, study habits)
25%

*Bloom (1976)

Quality of instruction represents 25 percent of your course grade, while affective student characteristics reflect the remaining 25 percent of your grade.

- *Intelligence* may be considered, for our purpose, how fast you can learn or relearn math concepts.
- *Cognitive entry skills* refer to how much math you knew before entering your current math course. For example, previous math knowledge can be based on previous math grades, the length of time since the last math course, the type of math courses completed (algebra I, algebra II, intermediate algebra), and placement test scores.
- *Quality of instruction* is concerned with the effectiveness of math instructors when presenting material to students in the classroom and math lab. This effectiveness depends on the course textbook, curriculum, teaching style, extra teaching aids (videos, audiocassettes), and other assistance. Quality of instruction may also include math lab resources, quality of tutoring based on tutor training and Supplemental Instruction.
- *Affective student characteristics* are characteristics you possess that affect your course grades — excluding how much math you knew before entering your math course. Some of these affective characteristics are anxiety, study skills, study attitudes, self-concept, motivation, and test-taking skills.

How what you know about math affects your grades

Poor math knowledge can cause low grades. A student placed in a math course that requires a more extensive math background than the student possesses will probably fail that course. Without the correct math background, you will fall behind and never catch up.

Placement Tests and Previous Course Grades

The math you need to know to enroll in a particular math course can be measured by a placement test (ACT, SAT) or by the grade received in the prerequisite math course. Some students are incorrectly placed in math courses by placement tests.

If, by the second class meeting, everything looks like Greek and you do not understand what is being explained, move to a lower-level course. In the lower-level math course you will have a better chance to understand the material and to pass the course. A false evaluation of math ability and knowledge can only lead to frustration, anxiety, and failure.

Requests by Students for Higher Placement

Some students try to encourage their instructors to move them to a higher-level math course because they believe they have received an inaccurate placement score. Many students try to avoid noncredit math courses, while other students do not want to repeat a course that they have previously failed. These moves can also lead to failure.

Some older students imagine that their math skills are just as good as when they completed their last math course, which was five to ten years ago. If they have not been practicing their math skills, they are just fooling themselves. Still other students believe they do not need the math skills obtained in a prerequisite math course to pass the next course. This is also incorrect thinking. Research indicates that students who were placed correctly in their prerequisite math course, and who subsequently failed it, will not pass the next math course.

What My Research Shows

I have conducted research on thousands of students who have either convinced their instructors to place them in higher-level math courses or have placed themselves in higher-level math courses. The results? These students failed their math courses many times before realizing they did not possess the prerequisite math knowledge needed to pass the course. Students who, without good reason, talk their instructors into moving them up a course level are setting themselves up to fail.

To be successful in a math course, you must have the appropriate math knowledge. If you think you may have difficulty passing a higher-level math course, you probably do not have an adequate math background. Even if you do pass the math course with a D or C, research indicates that you will most likely fail the next higher math course.

It is better to be conservative and pass a lower-level math course with an A or B instead of making a C or D in a higher-level math course and failing the next course at a higher level.

This is evident when many students repeat a higher-level math course up to five times before repeating the lower-level math course that was barely passed. After repeating the lower-level math course with an A or B, these students passed their higher-level math course.

How quality of math instruction affects your grades

Quality of instruction accounts for 25 percent of your grade. Quality of instruction includes such things as classroom atmosphere, instructor's teaching style, lab instruction, and textbook content and format. All of these "quality" factors can affect your ability to learn in the math classroom.

Interestingly enough, probably the most important "quality" variable is the compatibility of an instructor's teaching style with your learning style. You need to discover your learning style and compare it to the instructional style. Noncompatibility can be best solved by finding an instructor who better matches your learning style. However, if you cannot find an instructor to match your learning style, improving your math study skills and using the math lab or learning resource center (LRC) can compensate for most of the mismatch.

Use of the math lab or learning resource center (LRC) can dramatically improve the quality of instruction. With today's new technologies, students are able to select their best learning aids. These learning aids could be videotapes, CD-ROMs, computer programs, math study skills computer programs, math texts, and websites.

The quality of tutors is also a major factor in the math lab or learning resource center. A low student-to-tutor ratio and trained tutors are essential for good tutorial instruction. Otherwise, the result is just a math study hall with a few helpers.

The math textbook should be up to date with good examples and a solutions manual. This increases the amount of learning from the textbook compared to older, poorly designed texts.

The curriculum design affects the sequence of math courses, which could cause learning problems.

Some math courses have gaps between them that cause learning problems for all students.

How affective student characteristics affect your math grades

Affective student characteristics account for about 25 percent of your grade. These affective characteristics include math study skills, test anxiety, motivation, locus of control, learning style, and other variables that affect your personal ability to learn math.

Most students do not have this 25 percent of the grade in their favor. In fact, most students have never been taught *any* study skills, let alone how to study math, specifically. Students also do not know their best learning style, which means they may study ineffectively by using their least effective learning style. However, it is not your fault that you were not taught math study skills or made aware of your learning style.

By working to improve your affective characteristics, you will reap the benefits by learning math more effectively and receiving higher grades. Thousands of students have improved their math affective characteristics and thereby their grades by using this workbook.

Variable 2: _____

Variable 3: _____

Section Review

1. List and explain the three variables that contribute to your academic success.

   Variable 1: _____

2. How does your placement test or previous math course grade affect your math learning?

_____

_____

_____

3. How does the quality of math instructors affect your grade?

_____

_____

_____

_____

_____

4. How do your affective characteristics affect your grade?

_____

_____

_____

_____

_____

_____

How to determine your learning style

Research has shown that matching a student's best learning style with the instructional style improves learning. Research has also shown that students who understand their learning style can improve their learning

effectiveness. A learning disadvantage will occur for students who do not know or do not understand their learning style. Students should talk to their instructor or counselor about taking one or more learning style inventories.

Taking Stock of Your Learning Style

There are different types of learning styles assessments. One type of learning style assessment focuses on learning modalities, whereas other assessments focus on cognitive or environmental learning styles. We will focus on learning modalities and cognitive learning styles.

Learning modalities
(using your senses)

If it is available to you, take a learning style inventory that measures _learning modalities_. Learning modalities focus on the best way your brain receives information — that is, learning _visually_ (seeing), _auditorially_ (hearing), or _kinesthetically_ (touching).

If possible, take an inventory that offers math-learning modalities, such as _Learning Styles Inventory_ (Brown and Cooper, 1978). Other learning style inventories can measure learning modalities, but they focus on English- or reading-learning modes. Sometimes students' math-learning modes may be different from their language-learning modes. Your instructor can go to www.academicsuccess.com to learn more about using the Learning Styles Inventory.

If you cannot locate a modality-type inventory, then you may want to do a self-assessment on your math-learning modality. Do you learn math best by seeing it, by hearing it, or by having hands-on learning experiences? Hands-on learning means you need to touch or feel things to best learn about them. Group learning is when two to four students get together to discuss math problems and talk about the solutions. Individual learning is studying by yourself, using your book, CD or web resources. Do you learn best in groups or by yourself?

This is not the most accurate way to determine your learning modality; however, you may use this informal assessment until you have an opportunity to take a more formal modality learning style inventory.

Cognitive learning styles
(processing what you sense)

The second type of learning-styles inventory focuses on *cognitive learning styles*. Cognitive learning styles describe how you process information, once you have heard, seen or felt the information. Based on McCarthy (1981), you can process information four different ways: Innovative, Analytic, Common sense, or Dynamic. You can find information on the McCarthy inventory by going to www.AboutLearning.com.

If you do not have the opportunity to take a cognitive learning-style inventory, you may want to do a self-assessment of your cognitive learning style. Which of the following best describes you? Remember that you may not have all the characteristics mentioned for each learning style. Look for the style that best matches you.

Innovative learners solve problems or make decisions by personally relating to them and using their feelings. They are interested in people and like cultural events. They want to work towards peace and harmony. They work towards self-improvement. Some of their careers are in counseling, social work, personnel, humanities, social sciences, English and most fields dealing with people. They function through life by using their values.

They learn best by listening and discussing their ideas with others, while searching for the meaning. Their best learning activities are motivational stories, simulations, group discussion, journal writing, integrative lectures, group projects, and discussion lectures. They usually like subjective tests. They are students with good ideas, and they find unique solutions to problems. They ask the question, "Why do I need to learn this?" (McCarthy, 1981). In the math course they may ask for the underlying value of learning the course material as it relates to them. Their favorite question is, "Why?" Is this your cognitive learning style?

Analytic learners look for facts and ask experts for advice when solving problems. They want to know what the experts think and the research to back up the facts. In some cases they are more interested in ideas and concepts than people. They want to collect information before making logical decisions. If they are having difficulty with a decision, they will re-examine the facts or go to experts. They are very uncomfortable with subjective decisions. They generally major in math and the hard sciences; they enjoy careers involving research (McCarthey, 1981).

They learn by thinking about the concepts and solving problems by using logic. They learn best in the traditional lecture classroom by instructors giving them facts. Their best learning activities are lecture with visual aids, textbook reading assignments, problem solving and demonstration by the instructor (expert), independent research and gathering data. They usually like objective tests. They want to know, "What exactly is the content or skill?" (McCarthey, 1981). They learn best by taking exact and well organized notes and consider class discussion an interruption in learning. They usually do not like group learning and prefer studying individually. They usually major in areas like math, physics, chemistry and accounting. Their favorite question is "What?" Is this your learning style?

Common-sense learners solve problems by knowing how things work. They want to know how to use the material today and how it can be used in real life. They want to use facts to build concepts and then test the practicality of the concepts. In their discussions they get right to the point. They are skills oriented and like to tinker with things. They use their sensory experiences to help make decisions (McCarthy, 1981).

They learn by testing concepts in a practical way and make decisions by applying theories. They want to know practical ways to use theories and concepts. They need hands on experiences as part of their learning such as labs and manipulatives. Some additional learning activities are homework problems, field trips, computer-aided instruction, and lecture with demonstration, individual reports and computer simulations. They usually like objective tests (Harb, Terry, Hurt, & Williamson, 1981). They major in applied sciences, health fields, computer science, engineering and technologies. Their favorite question is "How?" Is this your learning style?

Dynamic learners solve problems by looking at hidden possibilities and processing information concretely. They like self-discovery. They like new things and can adapt to change. They can be very flexible. They can also reach correct conclusions and answers without knowing how they did it. They need to know what else can be done with this information. They seek to answer this question, "If I do this, what possibilities will it create?" (McCarthy, 1981).

They learn best by independent self-instruction and trial-and-error practice. They are flexible and will take risks. They function in life by acting out and testing experiences. Some of their best learning activities are Socratic lecture, role playing, open-ended laboratories, student lectures, outside projects, simulations and group problem solving (Harb, Terry, Hurt, & Williamson, 1981). They major in business, marketing and sales occupations. Their favorite question is, "What can this become?" Is this your learning style?

By now you may have identified your cognitive learning style. If not, take time to examine how you are learning and then revisit these descriptions. Also you may have recognized that innovative learners are the opposite of common sense learners and analytic learners are the opposite of dynamic learners. This does not mean that you cannot learn from someone teaching in the opposite learning style, but it will take more concentration and better study skills.

REMEMBER

It is important to recognize that students who do not have learning styles that fit easily with learning math can still learn math. Good study skills and awareness on the part of the math instructor and student can compensate a mismatch of modality/ learning styles. Working together to find strategies to help improve the teaching and learning is most productive for everyone.

Section Review

1. My learning modality is _____

   because I learn math best by _____

   _____

   _____

   _____

2. My cognitive learning style is _____

   because I process math information by _____

   _____

   _____

   _____

Assessing your math strengths and weaknesses

Students have learning strengths and weaknesses that affect math learning and their course grades. They must identify and understand these strengths and weaknesses to know which strengths support them and which weaknesses to improve upon. After reading this section, you will be asked to complete the Math-Learning Profile Sheet (Figure 2).

> **Examples**: Having good math study skills is a *positive* math-learning characteristic, whereas having high test anxiety is a *negative* math-learning characteristic.

Measuring Your Math Study Skills

Math study skills is the main part of the affective characteristics that represents about 25% of your math grade. You can determine your math study skills expertise by taking the free Math Study Skills Evaluation at www.academicsuccess.com, or you can do a self determination about your math study skills. If you take the online Math Study Skills Evaluation, the results will indicate your score. If you score below 70, you need to really focus on developing a set of study skills. A score between 70 and 80 means you have average study skills and a score above 90 means that you have good study skills. A self determination can be made by asking yourself if you have ever been taught math study skills. If you

have, then you probably have either good or excellent math study skills. If you have not, then you most likely need to develop good math study skills.

Scoring low on the study skills inventory is not the end of the world. In fact, it might indicate that your problems with learning math may be due to your study habits instead of math intelligence.

A low score on the math study skills evaluation means that most of your learning problems are because you have not been taught math study skills. This workbook can teach you math study skills. If you had previous poor math grades, it was probably due to lack of math study skills, and this is not your fault.

You can improve your math study skills and become a successful math student. If you have good math study skills, you can still improve them to make A's in math. If you have excellent math study skills, you can still learn some math study skills that can help you in future courses.

Measuring Your Math Anxiety Level

It is also important to assess any math anxiety you might have, so that you can start managing it immediately. If you took the online Math Study Skills Evaluation, your answer should indicate if you "seldom", "often" or "almost always" become anxious and forget important concepts during a test.

If you have medium or high math test anxiety, it can be reduced. This problem can be fixed if you follow the suggestions in the next chapter. This also means that some of your problems learning math (if you have any) are due to anxiety instead of your intellectual ability to learn math.

If you have high text anxiety, then you will probably need to use the *How to Reduce Test Anxiety* audio cassette tape. You can order this tape by going to www.academicsuccess.com. It is

Figure 2
Math-Learning Profile Sheet

After completing this profile, use your answers to select which areas you would like to improve first. If you have a serious problem with math or test anxiety, get help in this area immediately. Circle the X that best describes your belief.

	Strongly Disagree	Disagree	Agree	Strongly Agree
1. I have good math study skills.	X	X	X	X
2. Math tests don't make me anxious.	X	X	X	X
3. I have a strong basic knowledge of math to pass this class.	X	X	X	X
4. I have a positive attitude about learning math.	X	X	X	X
5. It has been less than a year since I took my last math class.	X	X	X	X
6. I have taken a math study skills course before this class.	X	X	X	X
7. I have good reading skills to be successful in this class.	X	X	X	X
8. I know if I improve my study habits, I can improve my math grades.	X	X	X	X

This informal survey is to help you decide what kind of shape you are in to do well in your course.

Which areas are strong and help you do well in a math class?

Which areas keep you from doing the best you can in a math class? How can you change them?

important that you find support and guidance in managing this anxiety immediately!

You can reduce and learn to manage test anxiety! It just takes work!

Measuring Previous Math Knowledge

Now it's time to measure your previous math knowledge. Your previous math knowledge is an excellent predictor of future math success. In fact, the best predictor of math success is your previous math grade (if taken within one year of your current course).

The measurement of your math knowledge is based on your previous math grade or your math placement test. If your college does not require a placement test, contact the chairperson of the math department to help you with your course placement. Selecting a math course yourself is like playing Russian roulette with five bullets in the chamber. Without help, you will probably commit mathematical suicide.

Visit your counselor or instructor to discuss your math placement test score. Ask your counselor or instructor how your placement score relates to the scores for the math course below and above your course.

Find out if your score barely placed you into your course, if your score was in the middle, or if your score almost put you into the next course. If your placement score barely put you into your current math course, you will have difficulty learning the math required in that course.

A barely passing placement score means that you are missing some of the math knowledge you require to be successful, and you will really struggle to pass the course.

A middle placement score indicates that you probably have most of the math knowledge required for the course; however, you still may have difficulty passing the course. A high placement score means you have most of the math knowledge needed to pass

and may have a good chance of passing the course. (Still, if you have poor math study skills and high test anxiety, you may not pass the course.)

Students who have low placement test scores need to improve their math knowledge as soon as possible. Students with middle placement scores should also consider improving their math knowledge.

You need to improve your math skills within the first few weeks of the semester. If you wait too long, you will become hopelessly lost. Suggestions for improving your math knowledge are included in a later section in this chapter.

If you are taking your second math course, use your previous math course grade to determine your math knowledge. Your previous course grade is the best predictor of success in your next math course. (However, this is only true if your last math course was taken within the last year.)

Students who made an A or B in their previous math course have a good chance of passing their next math course. However, students who made a C in their previous math course have a poor chance of passing their next math course. Students who made below a C or withdrew from the previous math course have virtually no chance of passing their next math course.

What level of math knowledge preparation do you possess right now for this course? If you have poor or good previous math knowledge, you need to build up those math concepts and skills as soon as possible. Go to your instructor and discuss how to build up your skills. Some suggestions are to review the chapters of the previous math text or ask your instructor about some websites that help students review math. Students with poor math knowledge have about three to four weeks to improve their math before it can become a major problem. Don't wait. Start now to review your math skills in order to improve your mathematics learning and grades. Additional information on improving your math knowledge is in the next section.

Determining Your Math Attitude

Math attitude can play a major part in being successful in your course. Some students who have a poor attitude toward math may not attend class as often as they should or may procrastinate in doing their homework. Other students avoid math until they *have* to take it. What is your math attitude — good, neutral, or negative? Remember, students with negative math attitudes can pass math. It depends on what you do with your attitude that counts. Don't let your negative attitude cause problems in your math course. Also, if you have a good math attitude, don't forget to study. A good math attitude can only get you so far!

The Amount of Time Since Your Last Math Course

Another indicator of math success can be the length of time since your last math course. The longer the time since your last math course, the less likely you will be to pass the current course. The exception to this rule is if you were practicing math while you were not taking math courses. Math is very easily forgotten if not practiced.

How long has it been since your last math course? If it has been two or more years since your last math class, your math may be similar to someone who has a low placement score or made a C or lower in the last math class. You can improve your math knowledge by following the same suggestions made for those students and applying the information in the next section (How to improve your math knowledge).

Previous Study Skills Courses or Training

General study skills training that does not focus on math may help some students in time management, reading techniques, learning styles, and overcoming procrastination. Even though math requires specific study techniques, you can learn many other helpful general study skills and personal habits that will help you in math classes. If you have had no study skills training, then this workbook will help you in your math and other courses. Students with no past study skills training should also take advantage of any other opportunities to learn about college success skills in addition to learning from this workbook. Contact your learning centers, instructors or counselors about study skills workshops and when they will be offered.

Previous/Current Required Reading Courses

College reading skills are also needed for success in math courses. This is especially true when you are trying to solve story or word problems.

At community colleges and some universities, there are three levels of reading courses. Usually, the first two levels are noncredit reading courses, and the third level is for college credit.

If you are enrolled in the lowest reading level, you will have difficulty reading and understanding the math text. Students who are enrolled in the second level may also have difficulty reading the text. Students who are not required to take a reading class or who are reading at grade level may still have some difficulty reading the math text.

REMEMBER

Math texts are not written like English or history texts. Even students with college-level reading skills may experience difficulty understanding their math text.

Students not at college level reading can pass math courses; however, if you need help in reading, get that help as soon as possible. Or, many learning or study skills centers help students with reading skills for specific courses. Work with a learning specialist to design a system for reading your math textbook. If you do not know your reading level, ask to take a reading test. If in doubt about your reading level, take a reading class.

Your Locus of Control

"Locus of control" is a concept that describes how much control you feel you have over both your life and your course grades. Some students feel they have a lot of control over their lives and learning, while other students feel they have very little control over their lives or grades. For example, when a student with external locus of control does poorly on a test, the blame is put on the instructor, when in reality the student did not study well enough. On the other hand, the student with internal locus of control admits that studying more and paying more attention in class would have helped. Students can improve their internal locus of control and be successful in mathematics at the same time.

Your Best Learning Modality

Learning modalities are best described by your preference in how you receive lecture information, either visually, auditory or kinesthetically (hands-on). As discussed, some of these modalities match your math instructor and some do not. Most math instructors teach through a visual mode, with the distant second modality being auditory.

Even if your learning style is different from your math instructor's, there is no excuse for not learning math. The awareness of your learning modality and cognitive style can help you become a better learner in and outside of the classroom. This information will be discussed in Chapter 6, "How to Remember What You Have Learned."

Your Cognitive Learning Style

Cognitive learning styles are best described as how information is processed once it is received by your brain. Most math instructors process information *analytically*, and they therefore teach math the same way. Using *common sense* is the second most popular way math instructors cognitively process information.

If you do not process information analytically or through common sense, then you may be at a learning disadvantage. However, improved math study skills can compensate for a mismatch in learning modes/ cognitive styles. Chapter 6 suggests different study skills that can help you learn math through your preferred learning modality and cognitive style.

Evaluating Your Findings

Now that you understand some of the general characteristics that help students be successful, complete the informal Math-Learning Profile Sheet (Figure 2).

When you look at your profile, think of it as similar to going to a fitness center for the first time. A professional trainer asks questions, runs you through equipment, analyzing your strengths and weaknesses, and then designs a program for you. You are doing the same thing here.

Section Review

1. List your strengths for learning math.

   _____

   _____

   _____

   _____

   _____

   _____

   _____

2. List three areas you will start improving right now.

   Area One: _____

   Area Two: _____

   Area Three: _____

3. How will you improve these areas?

   Area One: _____

   _____

   _____

   _____

   _____

   Area Two: _____

   _____

   _____

   _____

   _____

   Area Three: _____

   _____

   _____

   _____

   _____

   _____

How to improve your math knowledge

Instructors always operate on the premise that you finished your previous math course just last week; they do not wait for you to catch up on current material. It does not matter if your previous math course was a month ago or five years ago. Instructors expect you to know the previous course material — period.

Review Your Previous Math Course Material and Tests

There are several ways to improve your math knowledge. Review your previous math course material before attending your present math course. Look closely at your final exam to determine your weak areas. Work on your weak areas as soon as possible so they can become building blocks (instead of stumbling blocks) for your current course.

If it has been some time since your last math course, visit the math lab or learning resource center to locate review material. Ask the instructor if there are any computer programs that will assess your math skills to determine your strengths and weaknesses for your course. Review math videotapes on the math course below your level. Also review any computer software designed for the previous math course.

Another way to enhance learning is to review the previous math course text by taking all of the chapter review tests. If you score above 80 percent on one chapter review test, move on to the next chapter. A score below 80 percent means you need to work on that chapter before moving on to the next chapter. Get a tutor to help you with those chapters if you have trouble. Make sure you review all the chapters required in the previous course as soon as possible. If you wait more than two weeks to conclude this exercise, it may be too late to catch up (while learning new material at the same time).

Employ a Tutor

One last way to improve your cognitive entry skills is to employ a private tutor. If you have a history of not doing well in math courses, you may need to start tutorial sessions *the same week class begins.* This will give the tutor a better chance of helping you regain those old math skills.

You still need to work hard to relearn old math skills while continuing to learn the new material. If you wait four to five weeks to employ a tutor, it will probably be too late to catch up and do well or even pass the course.

REMEMBER

Tutorial sessions work best when the sessions begin during the first two weeks of a math course.

Schedule Math Courses "Back to Back"

Another way to maintain appropriate math knowledge is to take your math courses "back to back." It is better to take math courses every semester — even if you do not like math — so that you can maintain sequential (linear) learning.

I have known students who have made B's or C's in a math class and who then waited six months to a year to take the next math course. Inevitably, many failed. These students did not complete any preparatory math work before the math course and were lost after the second chapter. This is similar to having one semester of Spanish, not speaking it for a year, then visiting Spain and not understanding what is being said.

The only exception to taking math courses "back to back" is taking a six-week "kamikaze" math course (an ultracondensed version of a regular course), which should be avoided.

If you are one of the unfortunate many who are currently failing a math course, you need to ask yourself, "Am I currently learning any math or just becoming more confused?" If you are learning some math, stay in the course. If you are getting more confused, withdraw from the course. Improve your math knowledge prior to enrolling in a math course during the next semester.

Example: You have withdrawn from a math course after midterm due to low grades. Instead of waiting until next semester, attend a math lab or seek a tutor and learn Chapters 1, 2, and 3 *to perfection*. Also use this time to improve your math study skills. You will enter the same math course next semester with excellent math knowledge and study skills. In fact, you can make an A on the first test and complete the course with a high grade. Does this sound farfetched? It may, but I know hundreds of students who have used this learning procedure and passed their math course instead of failing it again and again.

Finding Your Best Instructor

Finding an instructor who best matches your learning style can be a difficult task. Your learning style is important; your learning style is how you best acquire information.

Example: Auditory learners do better when hearing the information over and over again instead of carefully reading the information. If an auditory learner is taught by a visual-style instructor who prefers that students read materials on their own and who prefers working problems instead of

continued

describing them, the mismatch could cause the student to do worse than if the student were taught by an auditory instructor.

Most students are placed in their first math course by an academic advisor. Usually, academic advisors know who are the most popular and least popular math instructors. However, advisors can be reluctant to discuss teacher popularity. And unfortunately, students may want the counselor to devise a course schedule based on the student's time limits instead of teacher selection.

To learn who are the best math instructors, ask the academic advisor which math instructor's classes fill up first. This does not place the academic advisor in the position of making a value judgment; neither does it guarantee the best instructor. But it will increase the odds in your favor.

Another way to acquire a good math instructor is to ask your friends about their current and previous math instructors. However, if another student says that an instructor is excellent, make sure your learning style matches your friend's learning style. Ask you friend, "Exactly what makes the instructor so good?" Then compare the answer with how you learn best. If you have a different learning style than your friend, look for another instructor, or ask another friend whose learning style more closely matches your own.

To obtain the most from an instructor, discover your best learning style and match it to the instructor's teaching style. Most learning centers or student personnel offices will have counselors who can measure and explain your learning style. Interview or observe the instructor while the instructor is teaching. This process is time consuming, but it is well worth the effort!

Once you have found your best instructor, *do not* change. Remain with the same instructor for every math class whenever possible.

REMEMBER

The first step in becoming a better math student is knowing your learning strengths and weaknesses. Now you can focus on what you need to improve.

Whether your learning styles are a natural fit for learning math or not, you still own the responsibility to be smart in selecting your study system for math class. You can be successful!

Section Review

1. Give four examples of how you can review math course materials.

Example 1: _____

_____

_____

_____

Example 2: _____

_____

_____

_____

Example 3: _____

_____

Example 4: _____

_____

_____

_____

_____

2. What is the advantage of taking math courses back to back?

_____

_____

_____

_____

3. Give two examples of how to find your best math instructor.

Example 1: _____

_____

_____

_____

Example 2: _____

_____

_____

_____

Chapter 2 Review

1. _____ _____ _____ refers to how much math you knew before entering your current math course.

2. _____ _____ _____ are the learning skills you possess that affect your grades.

3. Students who talk their instructors into moving up a course level usually _____ that course.

4. The _____ of tutors is a major part of the math lab or learning resource center.

5. Affective student characteristics account for about _____ percent of your grade.

6. _____ learners look for facts and ask experts for their advice.

7. Hands-on learners are called _____ learners.

8. Assessing your math-learning _____ and _____ will help you improve your studying and learning.

9. Students having previous difficulty in mathematics should start tutorial sessions the _____ week class begins.

10. Once you find your best math instructor _____ _____ change instructors for your next math course.

What is the most important information you learned from this chapter?

_____

_____

_____

_____

_____

_____

_____

_____

_____

_____

_____

How can you immediately use it?

_____

_____

_____

_____

_____

_____

_____

_____

_____

_____

Lecture Notes

3

How to Reduce Math Test Anxiety

In this chapter you will first learn about math anxiety and then explore test anxiety. They are different issues that can affect each other. It is best to learn about them as separate entities first.

Math anxiety is a common problem for many high school, college and university students. It is especially difficult for students in developmental courses who normally have more math anxiety than other students. However, there are students who are in higher level math courses that also struggle with this problem. It is very common for students to have anxiety only about math and not in their other subjects.

Most students think that math anxiety only affects them when taking a test, but it also affects other areas. It can affect the way you do your homework, learn in the classroom or through distance learning courses, and the way you choose a career. Students who have math anxiety may procrastinate in doing their homework or put off sitting down and completing an online lesson. This can lead to math failure. Students also select a major based on the amount of math that is required, which could lead to lower paying or dissatisfying careers. However, most students with math anxiety meet it face to face during the test, experiencing test anxiety as well.

Mild test anxiety can be a motivational factor in that it can make students properly prepare for a test. However, high test anxiety can cause major problems in both learning and test taking, as students avoid studying for the test when anxiety begins to afflict their thought processes. Reducing test anxiety is the key for many students to become successful in math. Such students need to learn the causes of test anxiety and how to reduce the test anxiety that affects their learning and grades.

Several techniques have proven helpful in reducing both math anxiety and math test anxiety. However, reducing them does not guarantee good math grades. It must be coupled with effective study skills and a desire to do well in math.

In this chapter, you will learn about math anxiety:

- definitions of math anxiety,
- types of math anxiety,
- causes of math anxiety, and
- effects of math anxiety on learning.

And you will learn about test anxiety:

- definitions of test anxiety

- why math tests create anxiety
- causes of test anxiety
- different types of test anxiety
- how to reduce test anxiety

Understanding math anxiety

Definition of Math Anxiety

Math anxiety is a relatively new concept in education. During the 1970s, certain educators began using the terms "mathophobia" and "mathemaphobia" as a possible cause for children's unwillingness to learn math. Math anxiety is an extreme emotional and/or physical reaction to a very negative attitude toward math. There is a strong relationship between low math confidence and high math test anxiety (Fox, 1977).

Math anxiety is the feeling of tension and anxiety that interferes with the manipulation of numbers and the solving of math problems during tests. (Richardson and Sulnn, 1972). Math anxiety is a state of panic, helplessness, paralysis, and mental disorganization that occurs in some students when they are required to solve math problems. This discomfort varies in intensity and is the outcome of numerous previous experiences students have had in their past learning situations (Tobias, 1986).

Today, math anxiety is accepted as one of the major problems students have in completing their math courses. It is real, but it can be overcome.

Types of Math Anxiety

Math anxiety can be divided into three separate anxieties: Math Test Anxiety, Numerical Anxiety and Abstraction Anxiety. Math Test Anxiety involves anticipation, completion and feedback of math tests. Numerical Anxiety refers to everyday situations requiring working with numbers and performing arithmetic calculations. Numerical anxiety can also include students who are trying to figure out the amount for a tip, thinking about mathematics, doing math homework or listening/seeing math instruction. Abstraction Anxiety deals with working with variables and mathematical concepts used to solve equations. Students can have all three math anxieties or only one anxiety. Most often, the students I have worked with have Math Test Anxiety and Abstraction Anxiety. These students don't have any anxiety working with numbers, but once they start learning algebra, they develop both conditions. This may have happened in high school or college.

The Causes of Math Anxiety

Since math anxiety is a learned condition, its causes are unique with each student, but they are all rooted in individuals' past experiences. Bad experiences in elementary school are one of the most common sources for students' math anxiety: coming in last in math races at the blackboard, watching a classmate next to them finish a problem twice as fast as they do, teachers saying, "That's okay. You just aren't good in math. You are better in English," or classmates and teachers calling them stupid. These words and experiences remain with people; they can still hear the words and eventually begin telling themselves the same thing. When these students walk into the classroom or open a math book, or take a test, these "mental tapes" play in their minds. When asked, many students indicated that they were made fun of when trying to solve math problems at the chalkboard. When they could not solve the problem, the teacher and/or students would call them "stupid."

Teacher and peer embarrassment and humiliation become the conditioning experience that causes some students' math anxiety. Over the years, this math anxiety is reinforced and even increases in magnitude. In fact, many math anxious students — now 30, 40 and 50 years old — *still* have extreme fear about working math problems on the board. One 56 year old indicated that he had a great deal of fear that the instructor would call him to the board. Even if he knew how to do the problem, displaying that knowledge to his peers was very difficult. Some students have said that they absolutely refused to go to the board.

Being embarrassed by family members can also cause math anxiety. According to students who have been interviewed on the matter, their parents tried to help them with math and this sometimes led to serious trauma. These students claim that the tutoring from their guardians, mainly their fathers, often resulted in scolding when they were not able to complete the problems. One student reported that his father hit him every time he got a multiplication problem wrong. Brothers and sisters can also tease one another about being dumb in math. This is particularly true of boys telling girls that they cannot do math. When people hear these statements enough times, they may start to believe them and associate these bad feelings with the word math. So, for students many years later just hearing the word "math" triggers a response of anxiety, consciously or unconsciously recalling the bad feelings, and becoming uneasy.

A good example of this is a student who I worked with who had completed her BS degree fifteen years ago at a college that did not require much math. She was returning to college to be an elementary school teacher, which required her to take math and a math placement test. As soon as I mentioned that she had to take math, she said, "I can't do math and I will have to wait a few days to get psychologically ready to take the math placement test." She indicated her old feeling of not being able

to do math rushed through her and she almost had an anxiety attack. This is an extreme case but a true example of math anxiety. In most cases math anxiety is not this bad, but it is disruptive enough to cause learning and testing problems.

If you have math anxiety, try to remember the first time you had anxiety about learning math. This doesn't include anxiety when taking a math test, which will be discussed later in this chapter. Was it in elementary, middle or high school? Can you recall a specific incident? Was it being called on in class or getting a poor grade on a homework assignment? Was it a parent saying he/she cannot do math and neither can you? If you can't remember a specific incident, then when was the last time you told yourself that you couldn't learn math? Today? Yesterday? A year ago?

Some students with math anxiety also have had positive experiences with learning math. Now, try to remember the first positive experience with math. Was it a teacher, parent, friend, or other person who praised you? Was it getting back a good grade on a homework assignment or test? Was it making a good grade in a math course? Think back to your last positive experience with math one more time. How did it make you feel? Those who don't have math anxiety still need to understand it, so they can help classmates that might have it and need support. Also, it is possible that they will develop it in the future.

One way to overcome math anxiety is to try to find out when and why it first occurred and how it is still affecting you. Right now, reread the questions in the previous two paragraphs and on separate paper, brainstorm your answers to the questions. You do not have to write complete sentences. Just record the information. If you think of other memories as you brainstorm, write those down too.

If some of the memories are tough ones, just make a short note of them right now. Don't dwell on them because they are in the past. This can be frustrating for some students, but it is worth working through it. You will realize that math anxiety is usually a result of events, not because of being "dumb" or other personal flaws.

If you have had a very positive experience in the past when you were studying math, brainstorm answers to the appropriate questions. You may want to keep in mind friends you know who struggle with math and think about the other questions.

This brainstorming will help you write your math autobiography assignment at the end of this chapter.

How Math Anxiety Affects Learning

Math Anxiety can cause learning problems in several ways. It can affect how you do your homework and your participation in the classroom and study groups. Let's first start by looking at how math anxiety could affect your homework. Students with high math anxiety may have difficulty starting or completing their homework. Doing homework reminds some students of their learning problems in math. More specifically, it reminds them of their previous math failures, which causes further anxiety. This anxiety can lead to total avoidance of homework or "approach-avoidance" behavior.

Total homework avoidance is called procrastination. The very thought of doing their homework causes these students anxiety, which causes them to put off tackling their homework. This makes them feel better for a short amount of time — *until test day.*

Math anxiety can also affect your classroom participation and learning. Usually students with math anxiety are afraid to speak out in class and ask questions. They remember that in the past they were made fun of when giving the wrong answer. They are also afraid of asking a question that others, including the teacher, will consider dumb. So they sit in class fearful of being asked a question, looking like they understand the lecture so they will not be

called on. They also take a lot of notes even though they don't understand them, in order to give the illusion of comprehension. If you are one of these students, these are hard habits to break. However, these habits may cause you to be unsuccessful in your math class. Here are some suggestions to break these habits:

1. Make an appointment to talk to your math instructor. Math instructors want you to talk to them. When I do my consulting around the country one of the major complaints I get from math instructors is that the students don't come and see them. Make an appointment to see your math instructor before the first major test to discuss your math history and to ask for learning suggestions. In most cases it is easier to talk to the instructor before you get your first grade.

2. Before class, ask the instructor to work one homework problem. You might want to write the problem on the board before the instructor arrives. This is less stressful because you are not asking the question in front of the whole class. In fact, one of my good friends, Dr. Mike Hamm, suggests that his students put the problems they do not know on the board before class. Other students go to the board and solve the problems. Dr. Hamm then solves the ones the students cannot do.

3. Prepare one question from your homework and ask it within the first 15 minutes of class. Instructors are more likely to answer questions in the first part of class when they have time instead of the end of class when time is running out.

4. Ask a question that you already know the answer. That way if the instructor starts to ask you a question about the problem, you will know the answer. This is good practice for asking those questions to which you don't know the answers.

5. Use email to send questions to your instructor. This way you can still get the answer with very little anxiety.

By using these suggestions you can reduce your math anxiety and learn more mathematics. A question unanswered could be a missed test question.

REMEMBER

The instructor's job is to answer your questions and you are paying for the course.

Math anxious students sometimes avoid doing additional math outside of the classroom. They avoid study groups and Supplemental Instruction. It is like asking a person with aquaphobia (fear of water) to take vacation at the beach. However, a person with aquaphobia can go to the beach and enjoy himself or herself and not get wet. In other words, students can still attend study groups and Supplemental Instruction and just listen. When they are ready to get their feet wet, they can ask a few questions. Don't let these great opportunities go by.

Math anxiety can affect how you learn mathematics. It can be overcome with your effort. You don't have to live in the past with your math fears. Today is a new day and you can change how math affects you. The next step is to understand how test anxiety can affect your demonstration of math knowledge.

Section Review

1. List and explain the three types of math anxiety.

   First Type: _____

   _____

   _____

   _____

   _____

   Second Type: _____

   _____

   _____

   _____

   _____

   Third Type: _____

   _____

   _____

   _____

   _____

2. List and explain two causes of math anxiety.

   First Cause: _____

   _____

   _____

   _____

   _____

   _____

   Second Cause: _____

   _____

   _____

continued

_____

_____

_____

3. List three ways math anxiety can cause learning
   problems.

   First Way: _____

   _____

   _____

   _____

   _____

   _____

   Second Way: _____

   _____

   _____

   _____

   _____

   _____

   Third Way: _____

   _____

   _____

   _____

   _____

   _____

4. List and explain three ways students can manage
   or avoid the affects of math anxiety.

   First Way: _____

_____

_____

_____

_____

_____

Second Way: _____

_____

_____

_____

_____

_____

Third Way: _____

_____

_____

_____

_____

_____

How to recognize test anxiety

Test anxiety has existed for as long as tests have been used to evaluate student performance. Because it is so common and because it has survived the test of time, test anxiety has been carefully studied over the last 50 years. Pioneering studies indicate that test anxiety generally leads to low test scores.

At the University of South Florida (Tampa), Dr. Charles Spielberger investigated the relationship between test anxiety and intellectual ability. The study results suggested that anxiety coupled with high ability can improve academic performance; but anxiety coupled with low or average ability can interfere with academic performance. That is:

Anxiety + high ability = improvement
Anxiety + low or average ability = no improvement

> **Example:** Students with average ability and low test anxiety had better performance and higher grades than did students with average ability and high test anxiety. However, there are students who make good grades, take calculus, and still have test anxiety.

Test anxiety is a *learned* response; a person is not born with it. An environmental situation brings about test anxiety. The good news is that because it is a learned response, it can be *unlearned*. Test anxiety is a special kind of general stress. General stress is considered "strained exertion," which can lead to physical and psychological problems.

Defining Test Anxiety

There are several definitions of test anxiety. One definition states, "Test anxiety is a conditioned emotional habit to either a single terrifying experience, recurring experience of high anxiety, or a continuous condition of anxiety" (Wolpe, 1958).

Another definition of test anxiety relates to the educational system. The educational system develops evaluations that measure one's mental performance, and this creates test anxiety. This definition suggests that test anxiety is the *anticipation* of some realistic or nonrealistic

situational threat (Cattell, 1966). The "test" can be a research paper, an oral report, work at the chalkboard, a multiple-choice exam, a written essay, or a math test.

Math test anxiety is a relatively new concept in education. *Ms.* magazine published "Math Anxiety: Why Is a Smart Girl Like You Counting on Your Fingers?" (Tobias, 1976) and coined the phrase *math anxiety*. During the 1970s, other educators began referring to *mathophobia* and *mathemaphobia* as a possible cause for children's unwillingness to learn math. Additional studies on the graduate level discovered that math anxiety was common among adults as well as children.

One of my students once described math test anxiety as "being in a burning house with no way out." No matter how you define it, math test anxiety is real, and it affects millions of students.

Why Math Tests Create Anxiety

It has been shown that math anxiety exists among many students who usually do not suffer from other tensions. Counselors at a major university reported that one-third of the students who enrolled in behavior therapy programs offered through counseling centers had problems with math anxiety (Suinn, 1988).

Educators know that math anxiety is common among college students and is more prevalent in women than in men. They also know that math anxiety frequently occurs in students with a poor high school math background. These students were found to have the greatest amount of anxiety.

Approximately half of the students in college prep math courses (designed for students with inadequate high school math background or low placement scores) could be considered to have math anxiety. However, math anxiety also occurs in students in high-level math courses, such as college algebra and calculus.

Educators investigating the relationship between anxiety and math have indicated that anxiety contributes to poor grades in math. They also found that simply *reducing* math test anxiety does not guarantee higher math grades. Students often have other problems that affect their math grades, such as poor study skills, poor test-taking skills, or poor class attendance.

Section Review

1. What is your personal definition of test anxiety?

   _____

   _____

   _____

   _____

   _____

2. What type of student does math anxiety affect most?

   _____

   _____

   _____

   _____

   _____

   _____

The causes of test anxiety

The causes of test anxiety can be different for each student, but they can be explained by seven basic concepts. See Figure 3 on the next page.

The causes of math test anxiety can be different for each student. It could possibly have first occurred in middle or high school. However, for many students it first occurs in college when passing tests is the only way to pass the course. Homework and extra credit in most college courses don't count toward your grade. Now students must have a passing average and in some cases must pass the departmental final exam. Additional pressure also exists because not passing algebra means you won't graduate and you might not get the job you want. As you can see, there are more reasons to have math test anxiety in college than in high school.

If you have math test anxiety, I want you to try to remember the first time it surfaced. Was it in middle school, high school or college? Can you recall a specific incident? Was it your first algebra test? Was it your first math test after being out of school for a long time? Was it after you decided to get serious about college? Was it a college algebra course that was required for your major? Was it an instructor who told you that if you don't pass the next math test you would fail the course? Was it when you needed to pass the next math test to pass the course so you could keep your financial aid? Was it your children asking you why you failed your last math test? If you cannot remember a specific incident when you had test anxiety, do you expect to have any major anxiety on the next test?

Students with math test anxiety also have had positive experiences with taking math tests. Now try to remember your first positive experience with taking a math test. Was it in middle school, high school or in college? Was it after studying many hours for the test? Now think back to your last positive experience with a math test. How did it make you feel?

Since we have already explored your experiences with taking math tests, let's look at some of the direct causes of your math test anxiety. If you do have test anxiety, what is the main cause? If you don't know, then review the seven Causes of Test Anxiety in Figure 3. Does one of these reasons fit you? If you don't have test anxiety, what would be a reason that could cause test anxiety?

Like you did before, brainstorm answers to the questions listed in the previous three paragraphs. You will also use this brainstorming to help you write your autobiography.

REMEMBER

For some students, just writing about their previous math history helps them.

Section Review

1. List and describe five causes of test anxiety.

First Cause: _____

_____

_____

_____

Figure 3
Causes of Test Anxiety

- Test anxiety can be a learned behavior resulting from the expectations of parents, teachers, or other significant people in the student's life.

- Test anxiety can be caused by the association between grades and a student's personal worth.

- Test anxiety develops from fear of alienating parents, family, or friends due to poor grades.

- Test anxiety can stem from a feeling of lack of control and an inability to change one's life situation.

- Test anxiety can be caused by a student's being embarrassed by the teacher or other students when trying to do math problems.

- Test anxiety can be caused by timed tests and the fear of not finishing the test, even if one can do all the problems.

- Test anxiety can be caused by being put in math courses above one's level of competence.

What are the cause(s) of *your* test anxiety?

Second Cause: _____

_____

_____

_____

_____

Third Cause: _____

_____

_____

_____

_____

Fourth Cause: _____

_____

_____

_____

_____

Fifth Cause: _____

_____

_____

_____

_____

2. Put into your own words your cause(s) of test anxiety. If you don't have test anxiety, list what you think is the major cause of test anxiety.

_____

_____

_____

_____

_____

_____

_____

_____

_____

_____

The different types of test anxiety

The two basic types of test anxiety are emotional (educators call this *somatic*) and worry (educators call this *cognitive*). Students with high test anxiety have *both* emotional and worry anxiety.

Signs of emotional anxiety are upset stomach, nausea, sweaty palms, pain in the neck, stiff shoulders, high blood pressure, rapid, shallow breathing, rapid heartbeat, or general feelings of nervousness. As anxiety increases, these feelings intensify. Some students even run to the bathroom to throw up or have diarrhea.

Even though these *feelings* are caused by anxiety, the physical response is real. These feelings and physical inconveniences can affect your concentration and your testing speed, and they can cause you to completely "draw a blank."

Worry anxiety causes the student to think about failing the test. These negative thoughts can happen either before or during the test. This negative "self-talk" causes students to focus on their anxiety instead of recalling math concepts.

The effects of test anxiety range from a "mental block" on a test to avoiding homework. One of the most common side effects of test anxiety is getting the test and immediately forgetting information that you know. Some students describe this event as having a "mental block," "going blank," or finding that the test looks like a foreign language.

After five or ten minutes into the test, some of these students can refocus on the test and start working the problems. They have, however, lost valuable time. For other students, anxiety persists throughout the test and they cannot recall the needed math information. It is only after they walk out the door that they can remember how to work the problems.

Sometimes math anxiety does not cause students to "go blank" but slows down their mental processing speed. This means it takes longer to recall formulas and concepts and to work problems. The result is frustration and loss of time, leading to *more* anxiety. Since, in most cases, math tests are *speed* tests (those in which you have a certain amount of time to complete the test), you may not have enough time to work all the problems or to check the answers if you have mentally slowed down. The result is a lower test score, because even though you know the material, you do not complete all of the questions before test time runs out.

Not using all of the time allotted for the test is another problem caused by test anxiety. Students know that they should use all of the test time to check their answers. In fact, math is one of the few subjects in which you can check test problems to find out if your work is correct. However, most students do not use all of the test time, and this results in lower test scores. Why does this happen?

Students with high test anxiety do not want to stay in the classroom. This is especially true of students whose test anxiety increases as the test progresses. The test anxiety gets so bad that they would rather leave early and receive a lower grade than stay in that "burning house."

Figure 4
The 12 Myths About Test Anxiety

Students are born with test anxiety.

Test anxiety is a mental illness.

Test anxiety cannot be reduced.

Any level of test anxiety is bad.

All students who are not prepared have test anxiety.

Students with test anxiety cannot learn math.

Students who are well prepared will not have test anxiety.

Very intelligent students and students taking high-level courses, such as calculus, do not have test anxiety.

Attending class and doing homework should reduce all test anxiety.

Being told to relax during a test will make a student relaxed.

Doing nothing about test anxiety will make it go away.

Reducing test anxiety will guarantee better grades.

Students have another reason for leaving the test early: the fear of what the instructor and other students will think about them for being the last one to hand in the test. These students refuse to be the last ones to finish the test because they think that the instructor or other students will think they are dumb. This is middle-school thinking, but the feelings are still real — no matter what the age of the student. These students do not realize that some students who turn in their tests first fail, whereas many students who turn in their tests last make A's and B's.

Another effect of test anxiety relates to completing homework assignments. Students who have high test anxiety may have difficulty starting or completing their math homework. Doing homework reminds some students of their learning problems in math. More specifically, it reminds them of their previous math failures, which causes further anxiety. This anxiety can lead to total homework avoidance or "approach-avoidance" behavior.

Total homework avoidance is called procrastination. The very thought of doing their homework causes these students anxiety, which causes them to put off tackling their homework. This makes them feel better for a short amount of time — *until test day.*

> **Example:** Some students begin their homework and work some problems successfully. They then get stuck on an problem that causes them anxiety, so they take a break. During their break the anxiety disappears until they start doing their homework again. Doing their homework causes more anxiety, which leads to another break. The breaks become more frequent.
>
> *continued*

> Finally, the student ends up taking one long break and not doing the homework. Quitting, to them, means *no more anxiety* until the next homework assignment.

The effects of math test anxiety can be different for each student. Students can have several of the mentioned characteristics that can interfere with math learning and test taking. However, there are certain myths about math that each student needs to know. Review Figure 4, The 12 Myths About Test Anxiety, to see which ones you believe. If you have test anxiety, which of the mentioned characteristics are true for you?

Section Review

1. List and describe the two basic types of test anxiety.

First Type: _____

_____

_____

_____

Second Type: _____

_____

_____

_____

2. List two reasons students leave the test room early instead of checking their answers.

First Reason: _____

_____

_____

Second Reason: _____

_____

_____

3. List 6 of the 12 myths about test anxiety that you most believed.

First Myth: _____

_____

_____

Second Myth: _____

_____

_____

Third Myth: _____

_____

_____

Fourth Myth: _____

_____

_____

Fifth Myth: _____

_____

_____

Sixth Myth: _____

_____

_____

How to reduce test anxiety

To reduce math test anxiety, you need to understand both the relaxation response and how negative self-talk undermines your abilities.

Relaxation Techniques

The relaxation response is any technique or procedure that helps you to become relaxed. It will take the place of an anxiety response. Someone simply telling you to relax or even telling yourself to relax, however, does little to reduce your test anxiety. There are both short-term and long-term relaxation response techniques that help control emotional (somatic) math test anxiety. These techniques will also help reduce worry (cognitive) anxiety. Effective *short-term* techniques include the tensing and differential relaxation method, the palming method, and deep breathing.

Short-Term Relaxation Techniques

The Tensing and Differential Relaxation Method

The tensing and differential relaxation method helps you relax by tensing and relaxing your muscles all at once. Follow these procedures while you are sitting at your desk before taking a test:

1. Put your feet flat on the floor.
2. With your hands, grab underneath the chair.
3. Push down with your feet and pull up on your chair at the same time for about five seconds.
4. Relax for five to ten seconds.
5. Repeat the procedure two or three times.
6. Relax all your muscles except the ones that are actually used to take the test.

The Palming Method

The palming method is a visualization procedure used to reduce test anxiety. While you are at your desk before or during a test, follow these procedures:

1. Close and cover your eyes using the center of the palms of your hands.
2. Prevent your hands from touching your eyes by resting the lower parts of your palms on your cheekbones and placing your fingers on your forehead. Your eyeballs must not be touched, rubbed, or handled in any way.
3. Think of some real or imaginary relaxing scene. Mentally visualize this scene. Picture the scene as if you were actually there, looking through your own eyes.
4. Visualize this relaxing scene for one to two minutes.
5. Open your eyes and wait about two minutes and repeat the visualization scene again. This time also imagine any sounds or smells that can enhance your scene. For example, if you are imagining a beach scene, hear waves on the beach and smell the salty air. This technique can also be completed without having your hands over your eyes by just closing your eyes.

Practice visualizing this scene several days before taking a test and the effectiveness of this relaxation procedure will improve.

Deep Breathing

Deep breathing is another short-term relaxation technique that can help you relax. Proper breathing is a way to reduce stress and decrease test anxiety. When breathing properly, enough oxygen gets into your bloodstream to nourish your body and mind. A lack of oxygen in your blood contributes to an anxiety state that makes it more difficult to react to stress. Proper deep breathing can help you control your test anxiety.

Deep breathing can replace the rapid, shallow breathing that sometimes accompanies test anxiety, or it can prevent test anxiety. Here are the steps to deep breathing:

1. Sit straight up in your chair in a good posture position.
2. Slowly inhale through your nose.
3. As you inhale, first fill the lower section of your lungs and work your way up to the upper part of your lungs.
4. Hold your breath for a few seconds.
5. Exhale slowly through your mouth.
6. Wait a few seconds and repeat the cycle.
7. Keep doing this exercise for four or five minutes. This should involve going through about ten breathing cycles. Remember to take two normal breaths between each cycle. If you start to feel light-headed during this exercise, stop for thirty – forty-five seconds and then start again.
8. Throughout the entire exercise, make sure you keep breathing smoothly and in a regular rhythm without gulping air or suddenly exhaling.
9. As an extra way to improve your relaxation, say "relax" or "be calm" to yourself as you exhale. This can start a conditioned response that can trigger relaxation when you repeat the words during anxious situations. As you keep practicing, this conditioned response will strengthen. Practice is the key to success.

You need to practice this breathing exercise for several weeks before using the technique during tests. If you don't practice this technique, then it will not work. After taking your first test keep doing the exercise several times a week to strengthen the relaxation response.

Side one of the audiocassette *How to Reduce Test Anxiety* (Nolting, 1986) further explains test anxiety and discusses these and other short-term relaxation response techniques. Short-term relaxation techniques can be learned quickly but are not as successful as the long-term relaxation technique. Short-term techniques are intended to be used while learning the long-term technique.

Long-Term Relaxation Techniques

The cue-controlled relaxation response technique is the best long-term relaxation technique. It is presented on side two of the audiocassette *How To Reduce Test Anxiety* (Nolting, 1986). Cue-controlled relaxation means you can induce your own relaxation based on repeating certain cue words to yourself. In essence, you are taught to relax and then silently repeat cue words, such as "I am relaxed."

After enough practice, you can relax during math tests. The cue-controlled relaxation technique has worked with thousands of students. For a better understanding of test anxiety and how to reduce it, listen to *How to Reduce Test Anxiety* (Nolting, 1986).

Negative Self-Talk

According to cognitive psychologists, self-talk is what we say to ourselves as a response to an event or situation. These dialogues determine our feelings about that event or situation. Sometimes we say it so quickly and automatically that we don't even hear ourselves. We then think it is the situation that causes the feeling, but in reality it is our interactions or thoughts about the experience that are controlling our emotions. This sequence of events can be represented by this timeline:

External Events
(math test)
↓
Interpretation of Events and Self-Talk
(how you feel about the test
and what you are telling yourself)
↓
Feelings and Emotions
(happy, glad, angry, mad, upset)

In most cases this means that you are responsible for how and what you feel. You can have positive self-talk or negative self-talk about taking a math test. Yes, some students have positive self-talk about math tests and see it as a challenge and something to accomplish that makes them feel good, while others see it as an upsetting event that leads to anger and anxiety. In other words, you are what you tell yourself.

Negative self-talk is a form of worry (cognitive) anxiety. This type of worrying can interfere with your test preparation and can keep you from concentrating on the test. Worrying can motivate you to study, but too much worrying may prevent you from studying at all.

Negative self-talk is defined as the negative statements you tell yourself before and during tests. Negative self-talk causes students to lose confidence and to give up on tests. Further, it can give you an inappropriate excuse for failing math and cause you to give up on learning math.

Examples of Negative Self-Talk

"No matter what I do, I will not pass this course."

"I failed this course last semester, and I will fail it again."

"I am no good at math, so why should I try?"

"I cannot do it; I cannot do the problems, and I am going to fail this test."

"I have forgotten how to do the problems, and I am going to fail."

"I am going to fail this test and never graduate."

"If I can't pass this test, I am too dumb to learn math and will flunk out."

Students who have test anxiety are prone to negative self talk. Test anxiety can be generated or heightened by repeatedly making statements to yourself that usually begin with "what if." For example, "What if I fail the test?" or "What if I fail this class again?" These "what if" statements generate more anxiety that can cause students to feel sick. These statements tell them to be anxious. Some other aspects of self-talk are:

- Self–talk can be in telegraphic form with short words or images.
- Self-talk can be illogical but at the time the person believes it.
- Negative self-talk can lead to avoidance like not taking the test or skipping classes.
- Negative self-talk can cause more anxiety.
- Negative self-talk can lead to depression and a feeling of helplessness.
- Negative self-talk is a bad habit that can be changed.

There are different types of negative self-talk. If you have negative self-talk, then review the types below and see which one fits you best. You may use a combination of them:

1. *The Critic* is the person inside us who is always trying to put us down. It constantly judges behaviors and finds fault even if it is not there. It jumps on any mistake and exaggerates it to cause more anxiety. The Critic puts us down for mistakes on the test and blames us for not controlling the anxiety. The Critic reminds us of previous comments from real people who have criticized us. It compares us to other students who are doing better in the class. It loves to say, "That was a stupid mistake!" or "You are a total disappointment. You can't pass this math class like everyone else can!" The Critic's goal is to promote low self-esteem.

2. *The Worrier* is the person inside us who looks at the worst-case scenario. It wants to scare us with the ideas of disasters and complete failure. When it sees the first sign of anxiety, it "blows it out of proportion" to the extent that we will not remember anything and totally fail the test. The Worrier creates more anxiety than normal. The Worrier anticipates the worst, underestimates our ability, and sees us not only failing the test but "failing life." The Worrier loves to ask "What if?" For example, "What if I fail the math test and don't graduate?" or "What if I can't control my anxiety and throw up in the math class?" The goal of the Worrier is to cause more anxiety so we will quit.

3. *The Victim* is the person inside us who wants us to feel helpless or hopeless. It wants us to believe that no matter what we do, we will not be successful in math. The Victim does not blame other events (poor schooling) or people (bad teachers) for math failures. It blames us. It dooms us and puts us into a learned helpless mode,

meaning that if we try to learn math, we will fail or if we don't try to learn math we will fail. So why try? The Victim likes to say, "I can't learn math." The goal of the Victim is to cause depression and make us not try.

4. *The Perfectionist* is similar to the Critic, but is the opposite of the Victim. It wants us to do the best we can and will guide us into doing better. It tells us that we are not studying enough for the math test and that a B is not good enough and that we must make an A. In fact, sometimes an A is not good enough unless it is a 100%. So, if we make a B on the test, the Perfectionist says, "A or B is just like making an F." The Perfectionist is the hard-driving part of us that wants the best but cannot stand mistakes or poor grades. It can drive us to mental and physical exhaustion to make that perfect grade. It is not interested in self-worth, just perfect grades. Students driven by the Perfectionist often drop a math course because they only have a B average. The Perfectionist loves to repeat, "I should have . . ." or "I must . . ." The goal of the Perfectionist is to cause chronic anxiety that leads to burnout.

Review these types of personalities to see which one may fit you best. We may have a little of each one in us, but what is the dominant one for you in math? Now we can look at how to stop these negative thoughts.

Managing Negative Self-Talk

Students need to change their negative self-talk to positive self-talk without making unrealistic statements.

Positive self-statements can improve your studying and test preparation. During tests, positive self-talk can build confidence and decrease your test anxiety. These positive statements (see examples) can help reduce your test anxiety and improve your grades. Some more examples of positive self-statements are on the cassette *How to Reduce Test Anxiety* (Nolting, 1986). Before the test, make up some positive statements to tell yourself.

There are several ways to counter and control negative self-talk. Students can replace the negative self-talk with positive statements and questions that make them think in a realistic way about the situation. Another way is to develop thought stopping techniques to reduce or eliminate the negative thoughts. Try each way or a combination to see what works best for you.

Countering self-talk involves writing down and practicing positive statements that can replace negative statements. Students can develop their own positive statements. Some rules for writing positive statements are:

1. Use the first person present tense, for example, "I can control my anxiety and relax myself."
2. Don't use negatives in the statement. For example, don't say, "I am not going to have anxiety." Instead say, "I will be calm during the test."
3. Have a positive and realistic belief in the statement. For example say, "I am going to pass this test," instead of saying, "I am going to make an A on this test" when you have not made an A on any of the tests.

The statements used to counter the negative thoughts can be based on the type of negative self-talk. The Critic who puts down by saying, "Your test anxiety is going to cause you to fail the test" can be countered with, "I have test anxiety, but I am learning to control it." The Worrier who says, "What if I fail the test" can be countered with "So what? I will do better on the next test." The Victim who thinks things are hopeless and says, "I will never be able to pass math" can be countered with, "I have studied differently for this test, and I can pass the math course." The Perfectionist who says, "I need to make an A on the test or I will drop out of school" can be countered with, "I don't need to make an A to please anyone. All I need is to pass the math course to get the career I want." These are some of the examples of counter statements that can control anxiety.

Examples of Positive Self-Talk

"I failed the course last semester, but I can now use my math study skills to pass this course."

"I went blank on the last test, but I now know how to reduce my test anxiety."

"I know my poor math skills are due to poor study skills, not my own ability, and since I am working on my study skills, my math skills will improve."

"I know that with hard work, I will pass math."

"I prepared for this test and will do the best I can. I will reduce my test anxiety and use the best test-taking procedures. I expect some problems will be difficult, but I will not get discouraged."

"I am solving problems and feel good about myself. I am not going to worry about that difficult problem; I am going to work the problems that I can do. I am going to use all the test time and check for careless errors. Even if I do not get the grade I want on this test, it is not the end of the world."

Thought-Stopping Techniques

Some students have difficulty stopping their negative self-talk. These students cannot just tell themselves to eliminate those thoughts. These students need to use a thought-stopping technique to overcome their worry and become relaxed.

Thought stopping involves focusing on the unwanted thoughts and, after a few seconds, suddenly stopping those thoughts by emptying your mind. Using the command "Stop!" or a loud noise like clapping your hands can effectively interrupt the negative self-talk. In a homework situation, you may be able to use a loud noise to stop your thoughts, but don't use it during a test.

To stop your thoughts in the classroom or during a test, silently shout to yourself "Stop!" or "Stop thinking about that." After your silent shout, either relax yourself or repeat one of your positive self-talk statements. You may have to shout to yourself several times during a test or while doing homework to control negative self-talk. After every shout, use a different relaxation technique/scene or positive self-talk statement.

Thought stopping works because it interrupts the worry response before it can cause high negative emotions. During that interruption, you can replace the negative self-talk with positive statements or relaxation. However, students with high worry anxiety should practice this technique three days to one week before taking a test. Contact your counselor if you have additional questions about the thought-stopping techniques.

Writing the Math Autobiography

An autobiography relates how you remember and feel about your past experiences. In addition, many autobiographies explore how these past feelings and experiences shape current life. While some people write autobiographies for others to learn about their lives, many people write private autobiographies in order to understand what is going on in their lives.

Use the Appendix as a guide to write your Math Autobiography. Your final product will be a typed (or handwritten if your instructor approves) paper that summarizes your experiences learning math, how you feel about these experiences, and how they shape your current perspective on learning math.

Section Review

1. Describe your best short-term relaxation technique.

   _____
   _____
   _____
   _____
   _____
   _____

2. You can use the palming method by closing your eyes and visualizing a scene without putting your hands to your face. Describe a very relaxing scene that you could visualize. Make sure to include some sounds and visual images in your scene.

   _____
   _____
   _____

   _____
   _____
   _____

3. Practice your relaxation scene for three to five minutes for the next five days. List the times and dates you practiced your scene.

   Date: _____ Time: _____
   Date: _____ Time: _____
   Date: _____ Time: _____
   Date: _____ Time: _____
   Date: _____ Time: _____

4. How does negative self-talk cause you to have text anxiety?

   _____
   _____
   _____
   _____
   _____
   _____

5. Make up three positive self-talk statements that are not listed in this text.

   Statement 1: _____
   _____

   Statement 2: _____
   _____

Statement 3: _____

_____

_____

6. Describe the thought-stopping technique in your own words.

_____

_____

_____

_____

_____

_____

_____

_____

_____

_____

_____

_____

_____

_____

_____

7. What words will you use as your silent shout?

_____

_____

_____

_____

8. What will you do after your silent shout?

_____

_____

_____

9. How does the thought-stopping technique work?

_____

Chapter 3 Review

1. Reducing math test anxiety _____ _____ guarantee good math grades.

2. Test anxiety is a _____ response; a person is not _____ with it.

3. Math test anxiety involves _____, _____, and _____ of math tests.

4. One cause of test anxiety is that a student goes to the board to work a problem but is called _____ when he or she cannot work the problem.

5. The two basic types of test anxiety are _____ and _____.

6. The effects of test anxiety range from a "_____ _____" on a test to _____ homework.

7. The _____ and _____ _____ method helps you relax by tensing and relaxing your muscles all at once.

8. The palming method is a _____ procedure used to reduce text anxiety.

9. Negative self-talk is a form of _____ anxiety.

10. Thought stopping works because it interrupts the _____ response before it can cause _____ emotions.

What is the most important information you learned from this chapter?

_____
_____
_____
_____
_____
_____
_____
_____
_____
_____
_____
_____

How can you immediately use it?

_____
_____
_____
_____
_____
_____
_____
_____
_____

Lecture Notes

How to Improve Your Listening and Note-Taking Skills

Listening and note-taking skills in a math class are very important, since most students either do not read the math text or have difficulty understanding it. In most of your other classes, if you do not understand the lecture, you can read the text and get almost all the information. In math class, however, the instructor can usually explain the textbook better than the students can read and understand it.

Students who do not have good listening skills or note-taking skills will be at a disadvantage in learning math. Most math understanding takes place in the classroom. Students must learn how to take advantage of learning math in the classroom by becoming effective listeners, calculator users, and notetakers.

In this chapter you will learn

- how to become an effective listener,
- how to become a good notetaker,
- when to take notes,
- the seven steps to math note-taking, and
- how to rework your notes.

How to become an effective listener

Becoming an effective listener is the foundation for good note-taking. You can become an effective listener using a set of skills that you can learn and practice. To become an effective listener, you must prepare yourself both physically and mentally.

Sitting in the Golden Triangle

The physical preparation for becoming an effective listener involves *where you sit* in the classroom. Sit in the best area to obtain high grades, that is, in "the golden triangle of success." The golden triangle of success begins with seats in the front row facing the instructor's desk (see Figure 5 on the next page).

Students seated in this area (especially in the front row) directly face the teacher and are most likely to pay attention to the lecture. This is a great seating location for visual learners. There is also less tendency for them to be distracted by activities outside the classroom or by students making noise within the classroom.

The middle seat in the back row is another point in the golden triangle for students to sit, especially those who are auditory (hearing) learners. You can hear the instructor better because the instructor's voice is projected to that point. This means that there is less chance of misunderstanding the instructor, and you can hear well enough to ask appropriate questions.

By sitting in the golden triangle of success, you can force yourself to pay more attention during class and be less distracted by other students. This is very important for math students because math instructors usually go over a point once and continue

on to the next point. If you miss that point in the lesson, then you could be lost for the remainder of the class.

This may seem obvious, but I know too many students who do the following if they have early morning classes. Roll out of bed, put on the closest clothes to them, brush their hair, maybe their teeth, and race to class. Don't do this. You won't wake up until the end of your class! Train yourself to be somewhere on campus to review your notes at least fifteen minutes before class starts and do some of the other strategies suggested in the next section. So, another physical way to be ready is to be awake and nourished!

Warming Up for Math Class

The mental preparation for note-taking involves "warming up" before class begins and becoming an active listener. Just as an athlete must warm up before a game begins, you must warm up before taking notes. Warm up by

- reviewing the previous day's notes,
- reviewing the reading material,
- reviewing the homework,
- preparing questions, and
- working one or two unassigned homework problems.

This mental warm up before the lecture allows you to refresh your memory and prepare pertinent questions, making it easier to learn the new lecture material.

Figure 5
The Golden Triangle of Success

Instructor

Visual Learners

Auditory Learners

Back of the Classroom

How to Become an Active Listener

Becoming an active listener is the second part of the mental preparation for note-taking.

Do not anticipate what the instructor is going to say or immediately judge the instructor's information before the point is made. This will distract you from learning the information.

> **Examples:** Watch the speaker, listen for main ideas, and nod your head or say to yourself, "I understand," when agreeing with the instructor.

Expend energy looking for interesting topics in the lecture. When the instructor discusses information that you need to know, immediately repeat it to yourself to begin the learning process.

You can practice this exercise by viewing math videotapes and repeating important information. This is an especially good learning technique for auditory learners.

Ask appropriate questions when you have them. Sometimes if you are so confused you do not know what question to ask, still let the instructor know you are confused. Also, mark your notes where you got lost so that if the instructor doesn't have time to answer your question, you can visit during office hours to get an answer.

REMEMBER

> Class time is an intense study period that should not be wasted.

Listening and Learning

Some students think that listening to the instructor and taking notes is a waste of valuable time. Students too often sit in class and use only a fraction of their learning ability. Class time should be considered a valuable study period where you can listen, take notes, and learn at the same time. One way to do this is by memorizing important facts when the instructor is talking about material you already know. Another technique is to repeat back to yourself the important concepts right after the instructor says them in class. Using class time to learn math is an efficient learning system.

Section Review

1. Explain how sitting in the golden triangle of success makes you a better listener.

    _____

    _____

    _____

    _____

    _____

2. List five ways you can warm up before math class begins.

    First Way: _____

    _____

    _____

    _____

Second Way: _____

_____

_____

Third Way: _____

_____

_____

Fourth Way: _____

_____

_____

Fifth Way: _____

_____

_____

3. Select your best warm-up process and try it before your next math class. What was your warm-up process, and how did it help you?

_____

_____

_____

_____

4. How can you listen and learn at the same time?

_____

_____

_____

_____

_____

_____

_____

_____

How to become a good notetaker

Becoming a good notetaker requires two basic strategies. One strategy is to be specific in detail. In other words, *copy* the problems down, step by step. The second strategy is to *understand* the general principles, general concepts, and general ideas.

Copying from the Board

While taking math notes, you need to copy each and every step of the problem even though you may already know every step of the problem. While in the classroom, you might understand each step, but a week later you might not remember how to do the problem unless all the steps were written down. In addition, as you write down each step, you are memorizing it. Make sure to copy every step for each problem written on the board.

There will be times when you will get lost while listening to the lecture. Nevertheless, you should keep taking notes, even though you do not understand the problem. This will provide you with a reference point for further study. Put a question mark (?) by those steps that you do not understand;

then, after class, review the steps you did not understand with the instructor, your tutor, or another student.

If you seem to think many other students are lost too, ask questions or let the instructor know.

Taking Notes

The goal of note-taking is to take the least amount of notes and get the greatest amount of information on your paper. This could be the opposite of what most instructors have told you. Some instructors tell you to take down everything. This is not necessarily a good note-taking system, since it is very difficult to take precise, specific notes while trying to understand the instructor.

What you need to develop is a note-taking system in which you write the least amount possible and get the most information down while still understanding what the instructor is saying.

Develop an Abbreviation List

To reduce the amount of written notes, an abbreviation system is needed. An abbreviation system is your own system of reducing long words to shorter versions that you still can understand. By writing less, you can listen more and have a better understanding of the material.

Example: When the instructor starts explaining the commutative property, you need to write "commutative property" out the first time. After that, use "COM." You should develop abbreviations for all the most commonly used words in math.

Figure 6
Abbreviations

E.G.	(for example)
CF.	(compare, remember in context)
N.B.	(note well, this is important)
\	(therefore)
∴	(because)
É	(implies, it follows from this)
>	(greater than)
<	(less than)
=	(equals, is the same)
¹	(does not equal, is not the same)
( )	(parentheses in the margin, around a sentence or group of sentences, indicates an important idea)
?	(used to indicate you do not understand the material)
O	(a circle around a word may indicate that you are not familiar with it; look it up)
E	(marks important materials likely to be used in an exam)
1, 2, 3,	(to indicate a series of facts)
D	(shows disagreement with statement or passage)
REF	(reference)
et al.	(and others)
bk	(book)
p	(page)
etc.	(and so forth)
V	(see)
VS	(see above)
SC	(namely)
SQ	(the following)
Com.	(commutative)
Dis.	(distributive)
APA	(associative property of addition)
AI	(additive inverse)
IPM	(identity property of multiplication)

Figure 6 on the previous page provides a list of abbreviations. Add your own abbreviations to this list. By using abbreviations as much as possible, you can obtain the same meaning from your notes and have more time to listen to the instructor.

When to take notes

To become a better notetaker, you must know when to take notes and when not to take notes. The instructor will give cues that indicate what material is important. These cues include

- presenting usual facts or ideas,
- writing on the board,
- summarizing,
- pausing,
- repeating statements,
- enumerating, such as "1, 2, 3" or "A, B, C,"
- working several examples of the same type of problem on the chalkboard,
- saying, "This is a tricky problem." Most students will miss it (for example, 5/0 is "undefined" instead of "zero").
- saying, "This is the most difficult step in the problem,"
- indicating that certain types of problems will be on the test, such as coin or age problems, and
- explaining bold-face words.

You must learn the cues your instructor gives to indicate important material. If you are in doubt about the importance of the class material, do not hesitate to ask the instructor about its importance.

While taking notes, you may become confused about math material. At that point, take as many notes as possible, and do not give up on note-taking. One thing you shouldn't do—bother the student next to you for an explanation. This can be annoying, and both of you will miss important information while you are talking.

As you take notes on confusing problem steps, leave extra space; then go back and fill in information that clarifies your misunderstanding of the steps in question. Ask your tutor or instructor for help with the uncompleted problem steps, and write down the reasons for each step in the space provided.

Another procedure to save time while taking notes is to stop writing complete sentences. Write your main thoughts in phrases. Phrases are easier to jot down and easier to memorize.

Section Review

1. Note-taking requires two basic strategies:

   _____ and

   _____.

2. List five abbreviations (and their meanings) that you use in your math class (one abbreviation and meaning per line).

   _____

   _____

   _____

   _____

   _____

3. List five cues that your instructor gives that indicate what material is important (one cue per line).

_____

_____

_____

_____

The seven steps to math note-taking

No matter what format you use to take notes, a good set of notes contains the following:

- examples of problems
- explanations of how to complete the problems
- appropriate math vocabulary and rules that are used for each problem

In addition, it is wise to make them as organized as possible. It may get on your nerves for a while, but when you see that organized notes save time when studying, you will be more willing to take the time to organize them.

Many successful math students, whether in developmental math courses or calculus, use a form of the three column note-taking system. It organizes the information, making it easy to review. While creating the three column system, students are also learning the math, so it is not busy work. It does take a few weeks to feel comfortable with it, but it is worth sticking with it. Many students have adopted this system and improved their studying time.

The seven steps lead you through several important stages in the learning process. Steps one through three help you *record the information* accurately, completely, and in an organized fashion. Steps four through seven help you *rehearse and understand the information*. These steps also help you learn to *recall the information*, helping you prepare for the test.

To prepare for taking notes this way, get a good chunk of paper ready. Here are some choices:

- Use regular paper, portrait or landscape style, and draw two lines to create three columns. They should be proportioned as follows:

- Some students purchase graph paper because it helps them keep numbers lined up correctly.
- Other students design the page on the computer and print out copies.
- Choose two columns next to each other to be the "Examples" and the "Explanations" columns. Label them as such. These are the two columns you will use all the time during class.
- Label the other column "Key Words and Rules." You might fill this column in during class but will probably complete it outside of class.

If you do this ahead of time, you do not have to frantically draw lines while the instructor starts talking. If you forget or run out of paper, it might be quicker to fold the paper into three columns.

Figure 7

Three Column Note-taking Method

Key Words/Rules	Example	Explanation
Solve linear equation	$4(x+4) + 3(x-4) = 2(x-1)$	Have to get x on one side of the $=$ and numbers on other side of the $=$
Distributive Property	$4x + 16 + 3x - 12 = 2x - 2$	Multiply numbers to the left of the ( ) with each variable and number in the ( ).
Commutative Property	$4x + 3x + 16 - 12 = 2x - 2$	Regroup numbers and variables.
Combine like terms	$7x + 4 = 2x - 2$	
Additive inverse	$7x - 2x + 4 = 2x - 2x - 2$ $5x + 4 = -2$	Subtract $2x$ from both sides in order to get variables all on the right of the $=$
Additive inverse	$5x + 4 - 4 = -2 - 4$ $5x = -6$	Subtract 4 from both sides to get numbers all on the right of the $=$
Mulitiplicative inverse	$\dfrac{5x}{5} = \dfrac{-6}{5}$	Divide both sides by 5 to get x by itself on the left side of the $=$
Simplify	$x = -1\ 1/5$	Put as a whole number and fraction.
	Insert new problem	

Draw a line dividing the end of one problem from the beginning of the next. Or, do one problem per page to give plenty of room for notes. Graph paper can help students keep numbers and words lined up. Turn the paper landscape.

During class you will use two columns most of the time—the "Example" and the "Explanations" columns. Sometimes, when instructors really want their students to use this method, they will organize their board work to reflect what the notes should look like. Most of the time, however, this does not happen.

The "Examples" column is where you will place the examples the instructor uses during class.

The "Explanations" column is where you put the instructor's verbal explanations of the examples. If the instructor tends to explain by just doing the problem without talking about it, ask questions that will force the instructor to explain the problem verbally.

When the instructor simply explains a vocabulary word or rule, write the vocabulary word or a phrase naming the rule in the 'Key Words and Rules" column. Write the explanation in the "Explanation" column. If the instructor provides an example, obviously it goes in the "Example" column. If no example is provided, ask for one. Then if your wish is not granted, look one up in the textbook after class.

Here are the "Seven Steps to Taking Math Notes." Remember, these are the basic steps. You can adapt them to work for you.

Step 1 *Record the instructor's example problems in the "Example" column.*

Step 2 *Record the reasons for each step in the "Explanations" column.* After completing a problem, draw a horizontal line, separating it from the next problem.

If you get lost during the lecture, place a question mark in this column and move on.

Step 3 *If possible, record the key words and rules in the other column.* Here is an opportunity for you to ask what rules are being used if the instructor is assuming the class knows them and you do not. Otherwise, you will need to refer to the textbook, work with a tutor, or use online resources after class to complete this column.

It is important to understand the vocabulary and rules –not just know how to do them!

Step 4 *In order to keep everything in your heads, not letting anything slip out, use the notes to review.* Cover up the "Examples" and "Explanations" columns and quiz yourself on the vocabulary and rules. Or, fold back the "Key Words and Rules" column, look at the examples and explanations and choose the correct key words and rules used to complete the examples.

Step 5 *Place a check mark by the key words and concepts that you did not know.*

Step 6 *Review the information that you checked until you understand and remember it.*

Step 7 *Develop a glossary of the math terms and rules that are difficult for you to remember.*

For a while you might spend more time organizing this system outside of class. Some students have to still depend on their current process, concentrating on getting the information down. Then they transfer the information into the three column method. However, it is best to learn how to do most of the process during math class.

Remember the learning modalities and learning styles? This process can use all of them!

- *Kinesthetic:* Writing out the notes
- *Auditory:* Saying information out loud
- *Visual:* Looking at the notes and thinking; color code each column with highlighters, making sure you do not smear the pencil writing
- *Group:* Working on the notes with one or two other people to make sure they are complete and accurate

• *Individual:* Reading them alone while waiting to meet someone.

Current research is discovering that using language (vocabulary, discussion) to learn math helps students to understand the concepts more deeply. Then completing many problems helps master the skills and increase the speed which is important for preparing for tests. These notes will help you understand the math so that it will stick with you. You will remember more information. Homework will help you master the skills and increase the speed at which you can complete problems.

A Math Glossary

The math glossary is created to define a math vocabulary in your own words. Since math is considered a foreign language, understanding the math vocabulary becomes the key to comprehending math. Creating a glossary for each chapter of your textbook will help you understand math.

Your glossary should include all words printed in bold face in the text and any words you do not understand. If you cannot explain the math vocabulary in your own words, ask your instructor or tutor for help. You may want to use the last pages in your notebook to develop a math glossary for each chapter in your textbook. Some students use note cards. Review your math glossary before each test.

Section Review

1. List and describe the seven steps to note-taking.

First Step: _____

_____

_____

_____

Second Step: _____

_____

_____

_____

Third Step: _____

_____

_____

_____

Fourth Step: _____

_____

_____

_____

Fifth Step: _____

_____

_____

_____

Sixth Step: _____

_____

_____

_____

Seventh Step: _____

_____

_____

_____

2. Make several copies of the modified three-column note-taking system, and take notes using this system in your next math class. Or develop the modified three-column note-taking system in your own notebook.

3. List two benefits of a math glossary.

Benefit 1: _____

_____

_____

_____

Benefit 2: _____

_____

_____

_____

4. From your current math chapter, list and define five words that you can put in your math glossary.

Word 1: _____

_____

_____

Word 2: _____

_____

_____

Word 3: _____

_____

_____

Word 4: _____

_____

_____

Word 5: _____

_____

_____

How to rework your notes

The note-taking system does not stop when you leave the classroom. As soon as possible after class, rework your notes. You can rework the notes between classes or as soon as you get home. By reworking your notes as soon as possible, you can decrease the amount of forgetting. This is an excellent procedure to transfer math information from short-term memory to long-term memory.

REMEMBER

Most forgetting occurs right after learning the material. You need to rework the notes as soon as possible. Waiting means that you probably will not understand what was written.

Here are important steps in reworking your notes:

Step 1   *Rewrite the material you cannot read or will not be able to understand a few weeks later.* If you do not rework your notes, you will be frustrated when studying for a test if you come across notes you cannot read. Another benefit of rewriting the notes is that you immediately learn the new material. Waiting means it will take more time to learn the material.

Step 2   *Fill in the gaps.* Most of the time, when you are listening to the lecture, you cannot write down everything. It is almost impossible to write down everything, even if you know shorthand. Locate the portions of your notes that are incomplete. Fill in the concepts that were left out. In the future, skip two or three lines in your notebook page for anticipated lecture gaps.

Step 3   *Add additional key words and rules in the keywords and rules column.* These key words or ideas were the ones not recorded during the lecture.

Example: You did not know that you should add the *opposite* of 18 to solve a particular problem, and you incorrectly added 18. Put additional important key words and ideas (such as "opposite" and "negative of") in the notes; these are the words that will improve your understanding of math.

Step 4   *Make a problem log on those problems that the teacher worked in class.* The problem log is a separate section of your notebook that contains a listing of the problems (without explanations — just problems) that your teacher worked in class. If your teacher chose those problems to work in class, you can bet that they are considered important. The problems in this log can be used as a practice test for the next exam. Your regular class notes will not only contain the solutions but also all the steps involved in arriving at those solutions and can be used as a reference when you take your practice test.

Write the problem on the left half of the page. Then save some room on the right side for writing out what the test instructions might be for the problem. Sometimes we forget what the test instructions are, and we miss points on the test by following the directions wrong. Skip several lines to complete the problem when you use the problem log as a practice test.

Step 5   *Make a calculator handbook and put in it your keystroke sequences.* The calculator handbook can be a spiral-bound set of note cards or a separate section of your notebook that holds only calculator-related information. Your handbook should also include an explanation of when that particular set of keystrokes is to be used.

Step 6   *Reflection and synthesis.* Once you have finished going over your notes, review the major points in your mind. Combine your new notes with your previous knowledge to have a better understanding of what you have learned today.

If you are a group learner and even if you are not, it helps to review and reflect with other students.

Use a Tape Recorder

If you have a math class during which you cannot get all the information down while listening to the lecture, ask your instructor about using a tape recorder. To ensure success, the tape recorder must have a tape counter and must be voice activated.

The tape counter displays a number indicating the amount of tape to which you have listened. When you find you are in an area of confusing information, write the beginning and ending tape counter number in the left margin of your notes. When reviewing your notes, the tape count number will be a reference point for obtaining information to work the problem. You can also reduce the time it takes to listen to the tape by using the pause button to stop the recording of unnecessary material.

Ask Questions

To obtain the most from a lecture, you must ask questions in class. By asking questions, you improve your understanding of the material and decrease your homework time. By *not* asking questions, you create for yourself unnecessary confusion during the remainder of the class period. Also, it is much easier to ask questions in class about potential homework problems than it is to spend hours trying to figure out the problems on your own at a later time.

If you are shy about asking questions in class, write down the questions and read them to your instructor. If the instructor seems confused about the questions, tell him or her that you will discuss the problem after class. To encourage yourself to ask questions, remember

- you have paid for the instructor's help,
- five other students probably have the same question,
- the instructor needs feedback on his or her teaching to help the class learn the material, and
- there is no such thing as a "stupid" question.

Record Each Problem Step

The final suggestion in note-taking is to record each step of every problem written or verbally explained. By recording each problem step, you begin *overlearning* how to work the problems. This will increase your problem-solving speed during future tests. If you get stuck on the homework, you will also have complete examples to review.

The major reason for recording every step of a problem is to understand how to do the problems while the instructor is explaining them instead of trying to remember unwritten steps. Although it may seem time consuming, it pays off during homework and test time.

Section Review

1. List and describe the six steps to reworking your notes.

First Step: _____

_____

_____

_____

Second Step: _____

_____

_____

_____

_____

Third Step: _____

_____

_____

_____

_____

Fourth Step: _____

_____

_____

_____

_____

Fifth Step: _____

_____

_____

_____

_____

Sixth Step: _____

_____

_____

_____

_____

2. How can using a tape recorder in class improve your learning?

_____

_____

_____

_____

3. List three reasons for asking questions in class.

First Reason: _____

_____

_____

_____

Second Reason: _____

_____

_____

_____

Third Reason: _____

_____

_____

_____

4. What is the major reason for recording each problem step?

_____

_____

_____

_____

_____

_____

Chapter 4 Review

1. Sitting in the golden triangle of success can help you pay more _____ during class.

2. Class time is an intense _____ _____ that should not be wasted.

3. Auditory learners should review the math videotape and _____ back the important information.

4. The goal of note-taking is to take the _____ amount of notes and get the _____ amount of information.

5. Using abbreviations can help you write _____ and listen _____.

6. Your _____ should include any words you don't know and their explanations.

7. Rework your notes to transfer information from _____-_____ memory to _____-_____ memory.

8. A _____ _____ is a separate section in your notebook that contains problems that your teacher worked in class.

9. Asking _____ in class will improve your understanding and decrease your homework time.

10. _____ how to work problems will increase your problem-solving speed and memory.

What is the most important information you learned from this chapter?

_____

_____

_____

_____

_____

_____

_____

_____

_____

_____

How can you immediately use it?

_____

_____

_____

_____

_____

_____

_____

_____

Lecture Notes

5

How to Improve Your
Reading, Homework, and Study Techniques

Reading a math textbook is more difficult than reading other textbooks. Math textbooks are written differently than your English or social science textbooks. Math textbooks contain condensed material and therefore take longer to read.

Mathematicians can reduce a page of writing to one paragraph using math formulas and symbols. To make sure you understood that same information, an English instructor would take that original page of writing and expand it into two pages. Mathematicians pride themselves on how little they can write and still cover the concept. This is one reason it may take you two to three times as long to read your math text as it would any other text.

REMEMBER

Reading your math text will take longer than reading your other texts.

Math students are expected to know how to do their homework; however, most math students do not have a homework system. Most students begin their homework by going directly to the problems and trying to work them. When they get stuck, they usually quit. This is not a good homework system.

A good homework system will improve your homework success and math learning at the same time.

In this chapter, you will learn

- how to read a math textbook,
- how to do your homework,
- how to solve word problems,
- how to work with a study buddy.

How to read
a math textbook

The way you read a math textbook is different from the traditional way students are taught to read textbooks in high school or college. Students are taught to read quickly or skim the material. If you do not understand a word, you are supposed to keep on reading. Instructors of other courses want students to continue to read so they can pick up the unknown words and their meanings from context.

This reading technique may work with your other classes, but using it in your math course will be totally confusing. By skipping some major concept words or bold-face words, you will not understand the math textbook or be able to do the homework. Reading a math textbook takes more time and concentration than reading your other textbooks.

If you have a reading problem, it would be wise to take a developmental reading course before taking math. This is especially true with reform math courses, where reading and writing are emphasized.

Reform math classes deal more with word problems than traditional math courses do. If you cannot take the developmental reading course before taking math, then take it during the same semester as the math course.

Ten Steps to Understanding Reading Materials

There are several appropriate steps in reading a math textbook:

Step 1   *Skim the assigned reading material.* Skim the material to get the general idea about the major topics. Read the chapter introduction and each section summary. You do not want to learn the material at this time; you simply want to get an overview of the assignment. Then think about similar math topics that you already know.

> **Example:** Skimming will allow you to see if problems presented in one chapter section are further explained in later chapter sections.

Step 2   *As you skim the chapter, circle (using*

pencil) *the new words that you do not understand.* If you do not understand these new words after reading the assignment, then ask the instructor for help. Skimming the reading assignments should take only five to ten minutes.

Step 3   *Put all your concentration into reading.* While reading the textbook, highlight the material that is important to you. However, do not highlight more than 50 percent of a page because the material is not being narrowed down enough for future study. Especially highlight the material that is also discussed in the lecture. Material discussed in both the textbook and lecture usually appears on the test. The purpose of highlighting is to emphasize the important material for future study. Do not skip reading assignments.

REMEMBER

> *Reading a math textbook is very difficult. It might take you half an hour to read and understand just one page.*

Step 4   *When you get to the examples, go through each step.* If the example skips any steps, make sure you write down each one of those skipped steps in the textbook for better understanding. Later on, when you go back and review, the steps are already filled in. You will understand how each step was completed. Also, by filling in the extra steps, you are starting to overlearn the material for better recall on future tests.

Step 5   *Mark the concepts and words that you do not know.* Maybe you marked them the first time while skimming. If you understand them now, erase the marks. If you do not understand the words or concepts, then reread the page or look them up in the glossary. Try not to read any further until you understand all the words and concepts.

Step 6 *If you do not clearly understand some words or concepts, add these words to the note-taking glossary in the back of your notebook.* Your glossary will contain the bold-face words that you do not understand. If you have difficulty understanding the bold-face words, ask the instructor for a better explanation. You should know all the words and concepts in your notebook's glossary before taking the test.

Step 7 *If you do not understand the material, follow these eight points, one after the other, until you do understand the material:*

Point 1 — Go back to the previous page and reread the information to maintain a train of thought.

Point 2 — Read ahead to the next page to discover if any additional information better explains the misunderstood material.

Point 3 — Locate and review any diagrams, examples, or rules that explain the misunderstood material.

Point 4 — Read the misunderstood paragraph(s) several times aloud to better understand their meaning.

Point 5 — Refer to your math notes for a better explanation of the misunderstood material.

Point 6 — Refer to another math textbook, computer software program, or videotape that expands the explanation of the misunderstood material.

Point 7 — Define exactly what you do not understand and call your study buddy for help.

Point 8 — Contact your math tutor or math instructor for help in understanding the material.

Step 8 *Recall important material.* Try to recall some of the important concepts and rules. If you have difficulty recalling this information, look at the highlighted areas of the text, bold print words or formulas. If you are a visual learner, write down these words several times on a separate sheet of paper. If you are an auditory learner, repeat these word several times out loud.

If you are a kinesthetic (hands on) learner, write out the formulas twice on two separate cards. Cut one of them up into puzzle pieces. Mix up the parts of the puzzle. Turn over the other formula card so you cannot see it. Put the puzzle back together. Do this several times.

For bold print words, write the bold print word on one card and its explanation twice on two other separate cards. Make another puzzle. Mix up the pieces of the explanation and put them back together.

Then add these words and formulas to your note-taking glossary. Recalling important material right after you read it improves you ability to remember that information.

Step 9   *Reflect on what you have read.* Combine what you already know with the new information that you just read. Think about how this new information enhances your math knowledge. Prepare questions for your instructor on the confusing information. Ask those questions at the next class meeting.

Step 10   *Write anticipated test questions.* Research has noted that students have about 80 percent accuracy in predicting test questions. Think about what is the most important concept you just read and what problems the instructor could give you that would test the knowledge of that concept. Make up three to five problems and add them to your problem log (this was part of your note-taking system). Indicate in the log that these are your questions, not the instructors.

By using this reading technique, you have narrowed down the important material to be learned. You have skimmed the textbook to get an overview of the assignment. You have carefully read the material and highlighted the important parts. You then added to your note-taking glossary unknown words or concepts.

REMEMBER

The highlighted material should be reviewed before doing the homework problems, and the glossary has to be learned 100 percent before taking the test.

How Reading Ahead Can Help

Reading ahead is another way to improve learning. If you read ahead, do not expect to understand everything. Read ahead two or three sections and put question marks (in pencil) by the material you do not understand.

When the instructor starts discussing that material, have your questions prepared and take good notes. Also, if the lecture is about to end, ask the instructor to explain the confusing material in the textbook. Reading ahead will take more time and effort, but it will better prepare you for the lectures.

How to Establish Study Period Goals

Before beginning your homework, establish goals for the study period. Do not just do the homework problems.

Ask yourself this question: "What am I going to do tonight to become more successful in math?"

By setting up short-term homework goals and reaching them, you will feel more confident about math. This also improves your self-esteem and helps you become a more internally motivated student. Set up homework tasks that you can complete. Be realistic.

Study period goals are set up either on a time-line basis or an item-line basis. Studying on a time-line basis is studying math for a certain amount of time.

Example: You may want to study math for an hour, then switch to another subject. You will study on a time-line basis.

Studying on an item-line basis means you will study your math until you have completed a certain number of homework problems.

Example: You might set a goal to study math until you have completed all the odd-numbered problems in the chapter review. The odd-numbered problems are the most important problems to work. These, in most texts, are answered in the answer section in the back of the book. Such problems provide the opportunity to recheck your work if you do not get the answer correct. Once you have completed these problems, do the even-numbered problems.

No matter what homework system you use, remember this important rule: Always finish a homework session by understanding a concept or doing a homework problem correctly.

Do not end a homework session with a problem you cannot complete. You will lose confidence, since all you will think about is the last problem you could not solve instead of the 50 problems you correctly solved. If you did quit on a problem you could not solve, return and rework problems you have done correctly.

REMEMBER

Do not end your study period with a problem you could not complete.

Section Review

1. List and describe the ten steps to reading a math textbook.

   First Step: _____
   _____
   _____
   _____

   Second Step: _____
   _____
   _____
   _____

   Third Step: _____
   _____
   _____
   _____

   Fourth Step: _____
   _____
   _____
   _____

   Fifth Step: _____
   _____
   _____
   _____

   Sixth Step: _____
   _____
   _____
   _____

Seventh Step: _____

_____

_____

_____

Eighth Step: _____

_____

_____

Ninth Step: _____

_____

_____

_____

Tenth Step: _____

_____

_____

_____

2. After trying each of the ten steps to reading your math text, make up, list, and try your own condensed version.

First Step: _____

_____

_____

Second Step: _____

_____

_____

Third Step: _____

_____

_____

Fourth Step: _____

_____

_____

Fifth Step: _____

_____

_____

Sixth Step: _____

_____

_____

Seventh Step: _____

_____

_____

_____

Eighth Step: _____

_____

_____

_____

Ninth Step: _____

_____

_____

_____

Tenth Step: _____

_____

_____

3. List two reasons why reading ahead can improve your learning.

First Reason: _____

_____

_____

_____

Second Reason: _____

_____

_____

_____

_____

4. Describe the two types of study period goals. Which one do you use?

First Goal: _____

_____

_____

Second Goal: _____

_____

_____

How to do your homework

Doing your homework can be frustrating or rewarding. Most students jump right into their homework, become frustrated, and stop studying. These students usually go directly to the math problems and start working them without any preparation. When they get stuck on one problem, they flip to the back of the text for the answer. Then they either try to work the problem backward to understand the problem steps or they just copy down the answer.

Other students go to the solution guide and just copy the steps. After getting stuck several times, these students will inevitably quit doing their homework assignment. Their homework becomes a frustrating experience, and they may even quit doing their math homework altogether.

10 Steps to Doing Your Homework

To improve your homework success and learning, refer to these ten steps:

Step 1 *Review the textbook material that relates to the homework.* A proper review will increase the chances of successfully completing your homework. If you get stuck on a problem, you will have a better chance of remembering the location of similar problems. If you do not review prior to doing your homework, you could get stuck and not know where to find help in the textbook.

REMEMBER

To be successful in learning the material and in completing homework assignments, you must first review your textbook.

Step 2 *Review your lecture notes that relate to the homework.* If you could not understand the explanation in the textbook on how to complete the homework assignment, then review your notes.

REMEMBER

Reviewing your notes will give you a better idea about how to complete your homework assignment.

Step 3   *Do your homework as neatly as possible.* Doing your homework neatly has several benefits. When approaching your instructor about problems with your homework, he or she will be able to understand your previous attempts to solve the problem. The instructor will easily locate the mistakes and show you how to correct the steps without having to decipher your handwriting. Another benefit is that when you review for midterm or final exams, you can quickly relearn the homework material without having to decipher your own writing.

REMEMBER

Neatly prepared homework can help you now and in the future.

Step 4   *When doing your homework, write down every step of the problem.* Even if you can do the step in your head, write it down anyway. This will increase the amount of homework time, but you are overlearning how to solve problems, which improves your memory. Doing every step is an easy way to memorize and understand the material. Another

advantage is that when you rework the problems you did wrong, it is easy to review each step to find the mistake.

REMEMBER

In the long run, doing every step of the homework will save you time and frustration.

Step 5   *Understand the reasons for each problem step and check your answers.* Do not get into the bad habit of memorizing how to do problems without knowing the reasons for each step. Many students are smart enough to memorize procedures required to complete a set of homework problems. However, when similar homework problems are presented on a test, the student cannot solve the problems. To avoid this dilemma, keep reminding yourself about the rules, laws, or properties used to solve problems.

Example: *Problem:* $2(a + 5) = 0$. What property allows you to change the equation to $2a + 10 = 0$? *Answer:* The distributive property.

Once you know the correct reason for going from one step to another in solving a math problem, you can solve any problem requiring that property. Students who simply memorize how to do problems instead of understanding the reasons for correctly working the steps will eventually fail their math course.

How to Check Your Answers

Checking your homework answers should be part of your homework duties. Checking your answers can improve your learning and help you prepare for tests.

Check the answers of the problems for which solutions are not given. This may be the even-numbered, or odd-numbered problems or the problems not answered in the solutions manual.

First, check your answer by estimating the correct answer.

> **Example:** If you are multiplying 2.234 by 5.102, remember that 2 times 5 is 10. The answer should be a little over 10.

You can also check your answers by substituting the answer back into the equation or doing the opposite function required to answer the question. The more answers you check, the faster you will become. This is very important because increasing your answer-checking speed can help you catch more careless errors on future tests.

Step 6   *If you do not understand how to do a problem, refer to the following points:*

Point 1 — Review the textbook material that relates to the problem.

Point 2 — Review the lecture notes that relate to the problem.

Point 3 — Review any similar problems, diagrams, examples, or rules that explain the misunderstood material.

Point 4 — Refer to another math textbook, solutions guide, math computer program software, or videotape to obtain a better understanding of the material.

Point 5 — Call your study buddy.

Point 6 — Skip the problem and contact your tutor or math instructor as soon as possible for help.

Step 7   *Always finish your homework by successfully completing problems.* Even if you get stuck, go back and successfully complete previous problems before quitting. You want to end your homework assignment with feelings of success.

Step 8   *After finishing your homework assignment, recall to yourself or write down the most important learned concepts.* Recalling this information will increase your ability to learn these new concepts. Additional information about step 8 will be presented later in this chapter.

Step 9   *Make up note cards containing hard-to-remember problems or concepts.* Note cards are an excellent way to review material for a test. More information on the use of note cards as learning tools is presented later in this chapter.

Step 10   *Do not fall behind.* As mentioned in Chapter 1, math is a sequential learning process. If you get behind, it is difficult to catch up because each topic builds on the next. It would be like going to Spanish class without learning the last set of vocabulary words. The teacher would be talking to you using the new vocabulary, but you would not understand what was being said.

Do Not Fall Behind

To keep up with your homework, it is necessary to complete the homework every school day and even on weekends. Doing your homework one-half hour each day for two days in a row is better than one hour every other day.

If you have to get behind in one of your courses, *make sure it is not math*. Fall behind in a course that does not have a sequential learning process, such as psychology or history. After using the ten steps to doing your homework, you may be able to combine two steps into one. Find your best combination of homework steps and use them.

REMEMBER

Getting behind in math homework is the fastest way to fail the course.

Section Review

1. List the ten steps to doing your math homework.

    First Step: _____

    _____

    _____

    Second Step: _____

    _____

    _____

Third Step: _____

_____

_____

Fourth Step: _____

_____

_____

Fifth Step: _____

_____

_____

_____

Sixth Step: _____

_____

_____

_____

Seventh Step: _____

_____

_____

_____

_____

Eighth Step: _____

_____

_____

_____

Ninth Step: _____

_____

_____

_____

Tenth Step: _____

_____

_____

2. After trying the ten steps to doing your math homework, make up, list, and try your own condensed version.

First Step: _____

_____

_____

Second Step: _____

_____

Third Step: _____

_____

_____

Fourth Step: _____

_____

_____

Fifth Step: _____

_____

_____

Sixth Step: _____

_____

_____

Seventh Step: _____

_____

_____

Eighth Step: _____

_____

_____

Ninth Step: _____

_____

_____

Tenth Step: _____

_____

_____

3. How will reviewing your notes or math text help you complete your homework more successfully?

_____

_____

_____

_____

_____

_____

_____

4. List two ways you can check your homework answers.

First Way: _____

_____

_____

_____

Second Way: _____

_____

_____

_____

5. What happens to students who fall behind in their homework?

_____

_____

_____

_____

_____

_____

How to solve word problems

The most difficult homework assignment for most math students is working story/word problems. Solving word problems requires excellent reading comprehension and translating skills.

Students often have difficulty substituting algebraic symbols and equations for English terms. But once an equation is written, it is usually easily solved. To help you solve word problems, follow these ten steps:

Step 1   *Read the problem three times.* Read the problem quickly the first time as a scanning procedure. As you are reading the problem the second time, answer these three questions:

1. *What is the problem asking me?* (This is usually at the end of the problem.)
2. *What is the problem telling me that is useful?* (Cross out unneeded information.)
3. *What is the problem implying?* (This is usually something you have been told to remember.) Read the problem a third time to check that you fully understand its meaning.

Step 2   *Draw a simple picture of the problem to make it more real to you* (e.g., a circle with an arrow can represent travel in any form — by train, by boat, by plane, by car, or by foot).

Step 3   *Make a table of information and leave a blank space for the information you are not told.*

Step 4   *Use as few unknowns in your table as possible.* If you can represent all the unknown information in terms of a single letter, do so! When using more than one unknown, use a letter that reminds you of that unknown. Then write down what your unknowns represent. This eliminates the problem of assigning the right answer to the wrong unknown. Remember that you have to create as many separate equations as you have unknowns.

Step 5   *Translate the English terms into an algebraic equation using the list of terms in Figures 8 and 9 on the next page.* Remember, the English terms are sometimes stated in a different order than the algebraic terms.

Step 6   *Immediately retranslate the equation, as you now have it, back into English.* The translation will not sound like a normal English phrase, but the meaning should be the same as the original problem. If the meaning is not the same, the equation is incorrect and needs to be rewritten. Rewrite the equation until it means the same as the English phrase.

Step 7   *Review the equation to see if it is similar to equations from your homework and if it makes sense.* Some formulas dealing with specific word problems may need to

Figure 8

Translating English Terms into Algebraic Symbols

Sum	+
Add	+
In addition	+
More than	+
Increased	+
In excess	+
Greater	+
Decreased by	-
Less than	-
Subtract	-
Difference	-
Diminished	-
Reduce	-
Remainder	-
Times as much	x
Percent of	x
Product	x
Interest on	x
Per	/
Divide	/
Quotient	/
Quantity	( )
Is	=
Was	=
Equal	=
Will be	=
Results	=
Greater than	>
Greater than or equal to	≥
Less than	<
Less than or equal to	≤

Figure 9

Translating English Words into Algebraic Expressions

English Words	Algebraic Expressions
Ten more than x	$x + 10$
A number added to 5	$5 + x$
A number increased by 13	$x + 13$
5 less than 10	$10 - 5$
A number decreased by 7	$x - 7$
Difference between x and 3	$x - 3$
Difference between 3 and x	$3 - x$
Twice a number	$2x$
Ten percent of x	$0.10x$
Ten times x	$10x$
Quotient of x and 3	$x/3$
Quotient of 3 and x	$3/x$
Five is three more than a number	$5 = x + 3$
The product of 2 and a number is 10	$2x = 10$
One-half a number is 10	$x/2 = 10$
Five times the sum of x and 2	$5(x + 2)$
Seven is greater than x	$7 > x$
Five times the difference of a number and 4	$5(x - 4)$
Ten subtracted from 10 times a number is that number plus 5	$10x - 10 = x + 5$
The sum of $5x$ and 10 is equal to the product of x and 15	$5x + 10 = 15x$
The sum of two consecutive integers	$(x) + (x + 1)$
The sum of two consecutive even integers	$(x) + (x + 2)$
The sum of two consecutive odd integers	$(x) + (x + 2)$

be rewritten. Distance problems, for example, may need to be written solving for each of the other variables in the formula. Distance = rate x time; therefore, time = distance/rate, and rate = distance/time. Usually, a distance problem will identify the specific variable to be solved.

Step 8   *Solve the equation using the rules of algebra.*

REMEMBER

Whatever is done to one side of the equation must be done to the other side of the equation. The unknown must end up on one side of the equation, by itself. If you have more than one unknown, then use the substitution or elimination method to solve the equations.

Step 9   *Look at your answer to see if it makes sense.*

Example: If tax was added to an item, it should cost more; if a discount was applied to an item, it should cost less. Is there more than one answer? Does your answer match the original question? Does your answer have the correct units?

Step 10   *Put your answer back into the original equation to see if it is correct.* If one side of the equation equals the other side of the equation, then you have the correct answer. If you do not have the correct answer, go back to step 5.

The most difficult part of solving word problems is translating part of a sentence into algebraic symbols and then into algebraic expressions. Review Figure 8 (Translating English Terms into Algebraic Symbols) and Figure 9 (Translating English Words into Algebraic Expressions), both on the previous page.

How to Recall What You Have Learned

After completing your homework problems, a good visual learning technique is to make note cards. Note cards are 3" by 5" index cards on which you place information that is difficult to learn or material you think will be on the test.

On the front of the note cards write a math problem or information that you need to know. Color-code the important information in red or blue. On the back of the note card, write how to work the problem or explain important information.

Example: If you are having difficulty remembering the rules for multiplying positive and negative numbers, you would write some examples on the front of the note card with the answers on the back.

Make note cards on important information you might forget. Every time you have five spare minutes, pull out your note cards and review them. You can glance at the front of the card, repeat to yourself the answer, and check yourself with the back of the card. If you are correct and know the information on a card, do not put it back in the deck. Mix up the cards you do not know and pick another card to test yourself on the information. Keep doing this until there are no cards left that you do not know.

If you are an auditory learner, then use the tape recorder like the note cards. Record the important information as you would on the front of the note card. Then leave a blank space on the recording before recording the answer. Play the tape back. When you hear the silence, put the tape on pause. Then say the answer out loud to yourself. Take the tape player off pause and see if you were correct.

Review What You Have Learned

After finishing your homework, close the textbook and try to remember what you have learned. Ask yourself these questions: "What major concepts did I learn tonight?" and "What test questions might the instructor ask on this material?"

Recall for about three to four minutes the major points of the assignment, especially the areas you had difficulty understanding. Write down questions for the instructor or tutor. Since most forgetting occurs right after learning the material, this short review will help you retain the new material.

Section Review

1. List the ten steps to solving word problems.

First Step: _____

_____

_____

Second Step: _____

_____

_____

Third Step: _____

_____

_____

Fourth Step: _____

_____

_____

Fifth Step: _____

_____

_____

Sixth Step: _____

_____

_____

Seventh Step: _____

_____

_____

Eighth Step: _____

_____

_____

Ninth Step: _____

_____

_____

Tenth Step: _____

_____

2. What is your best way to recall what you have learned?

_____

_____

_____

_____

_____

3. How does reviewing right after doing your homework help you remember?

_____

_____

_____

_____

_____

How to work with a study buddy

You need to have a study buddy when you miss class or when you do your homework. A study buddy is a friend or classmate who is taking the same course. You can find a study buddy by talking to your classmates or making friends in the math lab.

Try to find a study buddy who knows more about math than you do. Tell the class instructor that you are trying to find a study buddy and ask which students make the best grades. Meet with your study buddy several times a week to work on

problems and to discuss math. If you miss class, get the notes from your study buddy so you will not get behind.

Call your study buddy when you get stuck on your homework. You can solve math problems over the phone. Do not sit for half an hour or an hour trying to work one problem; that will destroy your confidence, waste valuable time, and possibly alienate your study buddy. Think how much you could have learned by trying the problem for 15 minutes and then calling your study buddy for help. Spend, at the maximum, 15 minutes on one problem before going on to the next problem or calling your study buddy.

REMEMBER

A study buddy can improve your learning while helping you complete the homework assignment. Just do not overuse your study buddy or expect that person to do your homework for you.

Section Review

1. How can you select your study buddy?

_____

_____

_____

_____

_____

2. How can a study buddy help you learn math?

_____

_____

_____

_____

_____

_____

3. Who is your study buddy?

_____

The benefits of study breaks

Psychologists have discovered that learning decreases if you do not take study breaks. Therefore, take a break after studying math for 45 minutes to one hour.

If you have studied for only 15 or 20 minutes and feel you are not retaining the information or your mind is wandering, take a break. If you continue to force yourself to study, you will not learn the material. After taking a break, return to studying.

If you still cannot study after taking a break, review your purpose for studying and your educational goals. Think about what is required to graduate. It will probably come down to the fact that you will have to pass math. Think about how studying math today will help you pass the next test; this will increase your chances of passing the course and of graduating.

Write on an index card three positive statements about yourself and three positive statements about studying. Look at this index card every time you have a study problem. Use every opportunity available to reinforce your study habits.

Section Review

1. How can study breaks improve your learning?

_____

_____

_____

_____

_____

_____

2. What do you do when you don't want to study? How do you get started?

_____

_____

_____

_____

_____

_____

Chapter 5 Review

1. Reading one page of a math text might take you _____ _____ _____.

2. After reading your textbooks, _____ important material can improve your ability to remember that information.

3. Write anticipated tests question in your _____ _____ and indicate that they are your questions.

4. Checking your homework answers can improve your _____ and help prepare you for a _____.

5. Matching:

(for students in pre-algebra class)

_____ Product   A. +

_____ Increased   B. -

_____ Sum   C. x

_____ Per   D. /

_____ Product   E. =

_____ Is

_____ Reduced

_____ Remainder

(for students taking elementary algebra or higher-level math course)

_____ Difference between   A. $10x$

x and 3   B. $x/3$

_____ Quotient of x and 3   C. $5(x-4)$

_____ Five times the sum   D. $5(x+2)$

of x and 2   E. $5 = x+3$

_____ A number added to 5   F. $5+x$

_____ Five less than 10   G. $10-5$

_____ Ten times x   H. $x-3$

_____ Five is three more than a number

_____ Five times the difference of a number and four

6. Make sure you finish your homework by working problems you _____ do.

7. Using _____ _____ is a visual way to recall what you learned in math.

8. Using a _____ _____ is an auditory way to recall what you learned in math.

9. Most_____ occurs right after learning; a short review will help you _____ the new material.

10. A _____ _____ can improve your learning while helping you complete the homework assignment.

What is the most important information you learned from this chapter?

_____
_____
_____
_____
_____
_____
_____

How can you immediately use it?

_____
_____
_____
_____
_____

6

How to Remember What You Have Learned

To understand the learning process, you must understand how your memory works. You learn by conditioning and thinking. But memorization is different from learning. For memorization, your brain must perform several tasks, including receiving the information, storing the information, and recalling the information.

By understanding how your memory works, you will be better able to learn at which point your memory is failing you. Most students usually experience memory trouble between the time their brain receives the information and the time the information is stored.

There are many techniques for learning information that can help you receive and store information without losing it in the process. Some of these techniques may be more successful than others, depending on how you learn best.

In this chapter, you will discover

- how you learn,
- how short-term memory affects what you remember,
- how working memory affects what you remember,
- how long-term memory/reasoning affects what you remember,
- how to use learning styles to improve memory,
- how to use memory techniques,
- how to develop practice tests, and
- how to use number sense.

How you learn

Educators tell us that learning is the process of "achieving competency." More simply put, it is how you become good at something. The three ways of learning are by conditioning, by thinking, and by a combination of conditioning and thinking.

Learning by Conditioning and Thinking

Conditioning is learning things with a maximum of physical and emotional reaction and a minimum of thinking.

Examples: Repeating the word *pi* to yourself and practicing where the

continued

symbol is found on a calculator are two forms of conditioned learning. You are learning using your voice and your eye-hand coordination (physical activities), and you are doing very little thinking.

Thinking is defined as learning with a maximum of thought and a minimum of emotional and physical reaction.

Example: Learning about *pi* by thinking is different from learning about it by conditioning. To learn "pi" by thinking, you would have to do the calculations necessary to result in the numeric value that the word *pi* represents. You are learning *using your mind* (thought activities), and you are using very little emotional or physical energy to learn *pi* in this way.

The most successful way to learn is to combine thinking and conditioning. The best learning combination is to learn by thinking first and conditioning second.

Learning by thinking means that you learn by

- *observing*,
- *processing*, and
- *understanding* the information.

How Your Memory Works

Memory is different from learning; it requires reception, storage, and retrieval of information. You

receive information through your five senses (*sensory input*): what you see, feel, hear, smell, and taste.

Examples: In math class, you will use your sense of *vision* both to watch the instructor demonstrate problems on the chalkboard and to read printed materials. You will use your sense of *hearing* to listen to the instructor and other students discuss the problems. Your sense of *touch* will be used to operate your calculator and to appreciate geometric shapes. In chemistry and other classes, however, you may additionally use your senses of *smell* and *taste* to identify substances.

The sensory register briefly holds an exact image or sound of each sensory experience until it can be processed. If the information is not processed immediately, it is forgotten. The sensory register helps us go from one situation to the next without cluttering up our minds with trivial information.

Processing the information involves placing it in either short-term memory, working memory, or long-term memory.

How short-term memory affects what you remember

Information that passes through the sensory register is stored in short-term memory. Remembering something for a short time is not hard to do for most students. By conscious effort, you can remember math laws, facts, and formulas received by the sensory register (your five senses) and put them in short-term memory. You can recognize them and register them in your mind as something to remember for a short time.

> **Example:** When you are studying math, you can tell yourself the distributive property is illustrated by $a(b + c) = ab + ac$. By deliberately telling yourself to remember that fact (by using conditioning — repeating or writing it again and again), you can remember it, at least for a while, because you have put it in short-term memory.

Psychologists have found that short-term memory cannot hold an unlimited amount of information. You may be able to use short-term memory to remember one phone number or a few formulas but not five phone numbers or ten formulas.

Items placed into short-term memory usually fade fast, as the name suggests.

> **Examples:** Looking up a telephone number in the directory, remembering it long enough to dial, then forgetting it immediately; learning the name of a person at a large party or in a class but forgetting it completely within a few seconds; cramming for a test and forgetting most of it before taking the test.

Short-term memory involves the ability to recall information immediately after it is given (without any interruptions). It is useful in helping you to concentrate on a few concepts at a time, but it is not the best way to learn. This is especially true of students who have attention problems or problems with short-term memory.

Students with short-term memory problems forget a math step as soon as the instructor explains it. To remember more facts or ideas and keep them in memory a longer period of time — especially at test time — use of a better system than short-term memory is required.

Changes in the way math is taught include the idea that conditioning (the most often used teaching form) leads only to short-term memory. Conditioning is often used for short-term recall of concept skills. The information learned in this way can only be stored for longer periods of time by thinking. Thinking takes place when applying the skills to working the problems.

How working memory affects what you remember

Working memory (or long-term retrieval) is that process in the brain that works on problems for a longer period of time than short-term memory. Working memory, then, offers an increase in the amount of *time* information is held in memory. (An increase in the *volume* of information that can be held requires long-term memory.)

Working memory is like the amount of RAM in a computer. Working memory uses the information (such as multiplication tables) recalled from long-term memory, along with new information to learn new concepts. It is the ability to think about and use many pieces of information at the same time. For instance, when you solve a linear equation you must use all the math that you learned in elementary school like addition, subtraction, and multiplication. You then have to add this to the new rules for linear equations.

Working memory can be compared to a mental workspace or an internal chalkboard. Just like a chalkboard, working memory has limited space, which can cause a "bottleneck" in learning. It involves the ability to recall information after learning has been consistently interrupted over a period of several minutes. Students with working memory problems may listen to a math lecture and understand each step as it is explained. When the instructor goes back to a previous step discussed several minutes prior, however, the student has difficulty explaining or remembering the reasons for the steps. These students have difficulty remembering series of steps long enough to understand the concept.

Working memory can go both ways in the memory process. First, it can go into long-term memory and abstract reasoning. Second, working memory can bring information out from long-term memory and abstract memory to use in learning new concepts. When learning a mathematical concept, working memory goes into abstract memory. When learning just facts such as the multiplication tables or definition of words, information goes directly into long-term memory. When solving math problems, in most cases, information is brought in from abstract reasoning and long-term memory into working memory. Students use their working memory to do their homework and when answering test questions. The amount of space in working memory is critical to answering test questions just like the amount of RAM is critical to running computer programs.

Example: In solving 26 x 32, you would put the intermediate products

continued

52 (*from 2 x 26*) and 780 (*from 30 x 26*—remember, 3 is in the 10s place, so make it 30) into working memory and add them together. The more automatic the multiplication, the less working memory you use. If you cannot remember your multiplication, you use up working memory trying to solve the multiplication problem. This leaves you with less working memory to solve the resulting addition problem.

How long-term memory/reasoning affects what you remember

Long-term memory is a storehouse of material that is retained for long periods of time. Working memory places information in long-term memory and long-term memory is recalled into working memory to solve problems. It is *not* a matter of trying harder and harder to remember more and more unrelated facts or ideas; it is a matter of organizing your short-term memories and working memories into meaningful information.

In most cases long-term memory is immeasurable. Students have so much room in their long-term memory that no one has measured its total capacity. Long-term memory also relates more to language skills rather than abstract skills. Students with good long-term memory and poor abstract skills can sometimes do well in every subject except math and the physical sciences. These students can use their long-term memory language skills by learning the vocabulary. By understanding the language of mathematics they can put into words how to solve math problems and recall these words during the test instead of depending mainly on their abstract memory. This information must be reviewed many times for it to get into long-term memory. This concept will be explored more in later chapters.

Reasoning or abstract memory is thinking about memories, comprehending their meanings and understanding their concepts. Abstract reasoning involves learning how the rules and laws apply to solving math problems. Students who understand the principals of a certain mathematical concept can use these principles to solve any similar problem. That is why the focus in mathematics is learning how to solve the problem based on concepts instead of just memorizing how to do a problem. In most cases you cannot memorize enough problems and hope they appear on the test. Without understanding the concept, the information cannot be transferred into abstract reasoning. Many students with excellent abstract reasoning major in math related careers.

The main problem most students face is converting information from working memory to long-term and/or abstract reasoning – and understanding it. To place information into long-term memory students must understand math vocabulary and practice problems. To place information into abstract reasoning students must understand the concept and remember it. In most cases student must use part of the long-term memory and abstract reasoning to solve math problems. It will depend on the skills in each area as to which one they use the best to solve math problems.

REMEMBER

Securing math information into long-term memory is not accomplished by just doing the homework — you must also understand it.

The Role of "Memory Output" in Testing

Memory output is what educators call a "retrieving process." It is necessary for verbal or written examinations. The retrieving process is used when answering questions in class, doing math homework or doing math tests. It is the method by which you recall information stored into long-term memory and by using abstract reasoning are able to place it into words or on paper. This retrieval process can come directly from long-term memory. For example, "What are whole numbers?" This is a fact question that comes from long-term memory. The retrieval process can also come from abstract reasoning. For example, "Write down the distributive property and substitute numbers to prove that it works." However, most math problems are solved through working memory by using information from both long-term memory and abstract reasoning. For example, solve: $3y - 10 = 9y + 21$. For this problem you are pulling in your number facts from long-term memory and the rules for solving equations from abstract reasoning.

Memory output can be blocked by three things:

1. insufficient processing of information into long-term memory,
2. test anxiety, or
3. poor test-taking skills.

If you did not completely place all of the information you learned into long-term memory, you may not be able to give complete answers on tests.

Test anxiety can decrease your ability to recall important information or it can cause you to block out information totally. Ways to decrease test anxiety were discussed in Chapter 3. Students who work on their test-taking skills can improve their memory output. Techniques to improve both your memory output and your test-taking skills will be explained in Chapter 7.

Understanding the Stages of Memory

Understanding the stages of memory will help you answer this common question about learning math: "How can I understand the procedures to solve a math problem one day and forget how to solve a similar problem two days later?"

There are three good answers to this question. First, after initially learning how to solve the problem, you did not *rehearse* the solving process enough for it to enter your long-term memory. Second, you did get the information into long-term memory, but the information was not *reviewed* frequently enough and was therefore forgotten. Third, you memorized how to work the problem but did not *understand* the concept.

REMEMBER

Locating where your memory breaks down and compensating for those weaknesses will improve your math learning.

Section Review

1. Label the boxes in Figure 10 on the facing page with stages of memory.

2. List, define and give examples for each stage of memory by explaining and giving examples of the following terms:

Stage One: _____
_____
_____
_____

Example: _____
_____
_____
_____

Stage Two: _____
_____
_____
_____

Example: _____
_____
_____
_____

Stage Three: _____
_____
_____
_____

continued on page 102

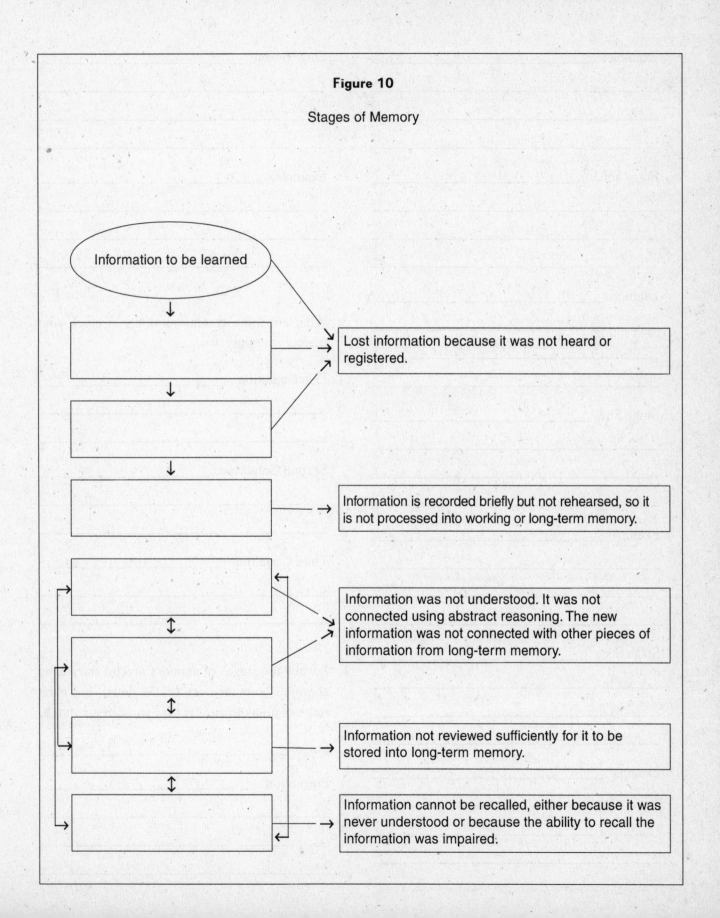

Figure 10

Stages of Memory

Information to be learned

Lost information because it was not heard or registered.

Information is recorded briefly but not rehearsed, so it is not processed into working or long-term memory.

Information was not understood. It was not connected using abstract reasoning. The new information was not connected with other pieces of information from long-term memory.

Information not reviewed sufficiently for it to be stored into long-term memory.

Information cannot be recalled, either because it was never understood or because the ability to recall the information was impaired.

Example: _____

_____

_____

_____

Stage Four: _____

_____

_____

_____

Example: _____

_____

_____

_____

Stage Five: _____

_____

_____

_____

Example: _____

_____

_____

_____

Stage Six: _____

_____

_____

_____

Example: _____

_____

_____

_____

Stage Seven: _____

_____

_____

_____

Example: _____

_____

_____

_____

3. What are three conditions that can block your memory output?

First Condition: _____

_____

_____

Second Condition: _____

_____

_____

Third Condition: _____

_____

_____

4. Review the stages of memory and list the stages at which *your* memory breaks down. For each stage of breakdown, say how you can prevent it.

Stage: _____

Prevention: _____

_____

_____

_____

Stage: _____

Prevention: _____

_____

_____

_____

Stage: _____

Prevention: _____

_____

_____

_____

5. What is the stage of memory you do well? Why?

Stage: _____

I do well because _____

_____

_____

_____

How to use learning styles to improve memory

There are many different techniques, which can help you store information in your long-term memory and reasoning. Using your learning sense or learning style and decreasing distraction while studying are very efficient ways to learn. Using your best *learning sense* (what educators call your "predominate learning modality") can improve how well you learn and enhance the transfer of knowledge into long-term memory/reasoning. Your learning senses are vision, hearing, touching, etc. Ask yourself if you learn best by watching (vision), listening (hearing), or touching (feeling).

Another helpful tool is the Learning Styles Inventory. It is the only inventory that has separate math learning style information. You can go to www.AcademicSuccess.com to learn more about the Learning Styles Inventory. Based on your preferred learning style practice those learning suggestions first.

REMEMBER

Learning styles are neither good nor bad and are based on genetics and environment. Knowing your best learning styles and using them effectively can dramatically improve your math learning and grades.

Visual (watching) Learner

Knowing that you are a visual math learner can help you select the memory technique that will work best for you. Repeatedly reading and writing down math materials being studied is the best way for a visual learner to study.

Based on the Learning Styles Inventory, students who learn math best by seeing it written are Visual Numerical Learners. If you are a visual numerical learner you will learn best by following the 10 suggestions in Figure 11, on the next page. Try as many of these suggestions as possible and then select and practice those that are most helpful.

A visual way to decrease distractions is by using the "my mind is full" concept. Imagine that your mind is completely filled with thoughts of

Figure 11
Visual Numerical Learners

These students learn math best by seeing it written. If you are a visual numerical learner, you may learn best by following these suggestions:

1. Studying a variety of written materials, such as additional handouts and math texts.

2. Playing games with and being involved in activities with visible printed number problems.

3. Using visually orientated computer programs, CD-ROMs, and math websites.

4. Reworking your notes using suggestions from this workbook.

5. Visualizing numbers and formulas, in detail.

6. Checking out videocassette tapes from the math lab or LRC.

7. Making 3x5 note (flash) cards, in color.

8. Using different colors of ink to emphasize different parts of the math formula.

9. Asking your tutor to *show you* how to do the problems instead of *telling you* how to do the problems.

10. Writing down each problem step the tutor tells you to do. Highlight the important steps or concepts, which cause you difficulty.

Auditory (hearing) Learner

If you are an *auditory learner* (one who learns best by hearing the information) then learning formulas is best accomplished by repeating them back to yourself, or recording them on a tape recorder and listening to them. Reading out loud is one of the best auditory ways to get important information into long-term memory. Stating facts and ideas out loud improves your ability to think and remember. If you cannot recite out loud, recite the material to yourself, emphasizing the key words.

Based on the Learning Styles Inventory students who learn math best by hearing it are Auditory Numerical Learners. If you are an auditory numerical learner you may learn best by following the 10 suggestions in Figure 12 on the next page. Try as many of these suggestions as possible and then select and practice those that are most helpful.

An auditory way to improve your concentration is by becoming aware of your distractions and telling yourself to concentrate. If you are in a location where talking out loud will cause a disturbance, mouth the words "start concentrating" as you say them in your mind. Your concentration periods should increase.

Tactile/Concrete (touching) Learner

A *tactile/concrete learner* needs to feel and touch the material to learn it. Tactile concrete learners, who are also called *kinesthetic* learners, tend to learn best when they can concretely manipulate the information to be learned. Unfortunately, most math instructors do not use this learning sense. As a result, students who depend heavily upon feeling and touching for learning will usually have the most difficulty developing effective

learning math, and other distracting thoughts cannot enter. Your mind has one-way input and output, which only responds to thinking about math when you are doing homework or studying.

Figure 12
Auditory Numerical Learners

1. Say numbers to yourself or moving your lips as you read problems.

2. Tape record your class and play it back while reading your notes.

3. Read aloud any written explanations.

4. Explain to your tutor how to work the math problem.

5. Make sure all important facts are spoken aloud with auditory repetition.

6. Remember important facts by auditory repetition.

7. Read math problems aloud and try solutions verbally and sub-verbally as you talk yourself through the problem.

8. Record directions to difficult math problems on audiotape and refer to them when solving a specific type of problem.

9. Have your tutor explain how to work the problems instead of just showing you how to solve them.

10. Record math laws and rules in your own words, by chapters, and listen to them every other day (auditory highlighting).

As mentioned before a tactile concrete learner will probably learn most efficiently by hands-on learning. For example, if you want to learn the FOIL method, you would take your fingers and trace the "face" to remember the steps. See Figure 14 (The FOIL Method). Also, learning is most effective when physical involvement with manipulation is combined with sight and sound. For example, as you trace the face you also say the words out loud.

Based on the Learning Styles Inventory, Tactile/Concrete Learners best learn math by manipulating the information that is to be taught. If you are a tactile/concrete learner, you may learn best by following the 10 suggestions in Figure 13. Try as many of these suggestions as possible and then select and practice the best suggestions that help. If you do not have these manipulatives or don't know how to use them, ask the math lab supervisor or instructor if they have any manipulative materials or models. If the math lab does not have any manipulative materials, you may have to ask for help to develop your own.

Tactile/concrete learners can also use graphing calculators to improve their learning. By entering the keystrokes it is easier to remember how to solve the problems. This practice is also an excellent way to remember how to solve the problem when using a calculator while taking a test.

Another way tactile/concrete learners can learn is to trace the graph with their fingers when it appears on the calculator. They should say out loud and trace every equation to "feel" how the graph changes when using different equations. For example, if you add 2 to one side of the equation, move your finger to where the graph changes and say out loud how much it moved.

A tactile/concrete way to improve your study concentration is by counting the number of distractions for each study session. Place a sheet of paper by your book when doing homework. When you catch yourself not concentrating put the letter

math learning techniques. This learning style creates a problem with math learning because math is more abstract than concrete. Also, most math instructors are visual abstract learners and have difficulty teaching math tactilely. Ask for the math instructors and tutors who give the most practical examples and who may even "act out" the math problems.

Figure 13
Tactile Concrete Learners

1. Cut up a paper plate to represent a fraction of a whole.

2. Fold up a piece of paper several times and cut along the fold marks to represent a fraction of a whole.

3. In order to understand math concepts, ask to be shown how to use qusinar rods or algebra tiles as manipulatives.

4. Try to use your hands and body to "act out" a solution. For example, you may "become" the car in a rate-and-distance word problem.

5. Obtain diagrams, objects or manipulatives and incorporate activities such as drawing and writing into your study time. You may also enhance your learning by doing some type of physical activity such as walking.

6. Try to get involved with at least one other student, tutor or instructor that uses manipulatives to help you learn math.

7. Ask to use the Hands-on Equations Learning System using manipulatives to learn basic algebra. You can go to their Web site (www.Borenson.com) to learn more about this system and other systems to help you learn math.

8. Go to one of the "learning stores" usually in your local mall to see if they have manipulatives.

9. Go to a K-12 learning resource store to see if they have manipulatives, such as magnetic boards with letters and numbers.

10. Talk to the coordinator of students with disabilities to see if they use manipulatives when tutoring their students.

Figure 14
The FOIL Method

F (a) (c)
O (a) (d)
I (b) (c)
L (b) (d)

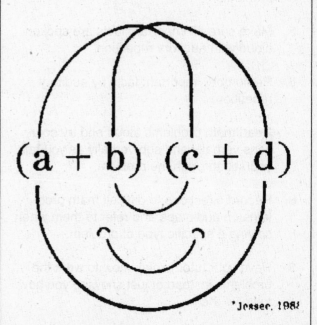

$(a + b)$ $(c + d)$

Jesser, 1988

FOIL is used to remember the procedure to multiply two binomials. The letters in FOIL stand for First, Outside, Inside and Last. To use FOIL, multiply the following

- the First terms ((a) (c)),
- the Outside terms ((a) (d)),
- the Inside terms ((b) (c)),
- the Last terms ((b) (d)).

To learn FOIL, trace your finger along the FOIL route.

"C" on the sheet of paper. This will remind you to concentrate and get back to work. After each study period, count up the number of "C's" and watch the number decrease.

Social Individual Learner

If you are a social individual learner, learning math may best be done individually. You may learn best by yourself, working with computer programs and being individually tutored. In some cases, social individuals may have to meet in groups to develop practice tests but leave socializing to a minimum. If you are a social individual learner and visual learner, using the computer may be one of the best learning tools available. If you are a social individual learner, based on the Learning Styles Inventory, you may learn best by following the eight suggestions in Figure 15. Try as many of these suggestions as possible and then select those that are most helpful.

A problem that a social individual learner may encounter is working too long on a problem for which they could have received help. Social individual learners must understand that getting help is okay, especially if it saves study time and makes them more study efficient.

Social Group Learners

If you are a social group learner (one who best learns in a group) then learning math may best be done in study groups and in math classes that have collaborative learning (group learning). Social group learners may learn best by discussing information. They can usually develop their own study groups and discuss how to solve problems over the phone. If you are a social group learner and an auditory learner then you definitely learn best by talking to people. If you are a social group learner,

---

Figure 15
Social Individual Learners

1. Study math, English or other subjects alone.

2. Utilize videocassette tapes or auditory tapes to learn by yourself.

3. Prepare individual questions for your tutor or instructor.

4. Obtain individual help from the math lab or hire your own tutor.

5. Set up a study schedule and study area so other people will not bother you.

6. Study in the library or in some other private, quiet place.

7. Use group study times only as a way to ask questions, obtain information and take pre-tests on your subject material.

8. Use math websites to help you learn.

---

based on the Learning Styles Inventory, you may learn best by following the eight suggestions in Figure 16, on the following page. Try as many of these suggestions as possible and then select and practice those that are most helpful.

A learning problem that a social group learner may have is talking too much about other subjects when in a study group. This is called being off task. You may want to have a student serve as a discussion monitor to let the students know when they need to get back on task. Also, social group learners need to know that they still must study math individually to be successful. During this individual study session prepare questions for the group.

Figure 16
Social Group Learners

1. Study math, English or your other subjects in a study group.

2. Sign up for math course sections which use cooperative learning (learning in small groups).

3. Review your notes with a small group.

4. Obtain help in the math lab or other labs where you can work in group situations.

5. Watch math videocassette tapes with a group and discuss the subject matter.

6. Listen to audiocassette tapes on the lecture and discuss them with the group.

7. Obtain several "study buddies" so you can discuss with them the steps to solving math problems.

8. Form a study group. Each member brings 10 test questions with explanations on the back. The group completes all the test questions.

Multiple Senses

If you have difficulty learning material from one sense (learning style), you might want to try learning material through two or three senses. Involving two or more senses in learning improves your learning and remembering. Review the figures in this section on using your learning styles and whenever possible combine the learning styles.

If your primary sense is visual and your secondary sense is auditory, you may want to write down equations while saying them out loud to yourself. Writing and reciting the material at the same time combines visual, auditory, and some tactile/concrete styles of learning.

Studying with a pen or highlighter is a visual as well as a tactile/concrete way to improve your concentration. Placing the pen or highlighter in your hand and using it will force you to concentrate more on what you are reading. After you write and recite the material back to yourself, do it five or ten more times to over learn it.

Section Review

1. List five ways visual learners can improve their memory:

First Way: _____

_____

Second Way: _____

_____

Third Way: _____

_____

Fourth Way: _____

_____

Fifth Way: _____

_____

2. List five ways auditory learners can improve their memory:

First Way: _____

_____

Second Way: _____

_____

Third Way: _____

_____

Fourth Way: _____

_____

Fifth Way: _____

_____

3. List five ways tactile/concrete learners can improve their memory:

First Way: _____

_____

Second Way: _____

_____

Third Way: _____

_____

Fourth Way: _____

_____

Fifth Way: _____

_____

4. List three ways social individual learners can improve their memory:

First Way: _____

_____

Second Way: _____

Third Way: _____

_____

5. List three ways social group learners can improve their memory:

First Way: _____

_____

Second Way: _____

_____

Third Way: _____

_____

6. What is your best modality learning style for math? Visual, auditory or tactile/concrete?

_____

7. What is your second best modality learning style for math? _____

8. Are you more of an individual or group learner? _____

Why? _____

_____

9. How can you use multiple senses to improve your memory?

_____

_____

_____

10. What is your best combination of learning styles to learn math? _____

_____

Give an example of how you can combine your learning styles to learn a math concept (if you cannot think of a concept ask your instructor for one).

_____

_____

_____

_____

_____

_____

How to use memory techniques

There are many different techniques that can help store information in your long-term memory: a positive attitude about studying, selective learning, organization, visual imagery, mnemonic devices, and acronyms.

A Good Study/Math Attitude

A positive attitude about studying will help you concentrate and improve your retention of information. This means you should have at least a neutral math attitude (you neither like nor dislike it) and reserve the right to actually learn to like math. If you still don't like math, just pretend that you do while studying it. View studying as an opportunity to learn rather than an unpleasant task. Tell yourself that you can learn the material, and it will help you pass the course and graduate.

Be a Selective Learner

Being selective in your math learning will improve your memory. Prioritize the materials you are studying; decide which facts you need to know and which ones you can ignore. Narrow down information into laws and principles that can be generalized. Learn the laws and principles 100 percent.

Example: If you have been given a list of math principles and laws to learn for a test, put each one on an index card. As you go through them, create two piles: a "I already know this" pile and a "I don't know this" pile. Then study *only* the "I don't know this" pile. Study the "I don't know this" pile until it is completely memorized and understood.

Become an Organizer

Organizing your math material into idea/fact clusters or groups will help you learn and memorize it. Grouping similar material in a problem log or calculator log is an example of categorizing information. Do not learn isolated facts; always try to connect them to other similar material.

Use Visual Imagery

Using mental pictures or diagrams to help you learn the material is especially helpful for the visual learners and those who are right-hemisphere dominant (who tend to learn best by visual and spatial methods). Mental pictures and actual diagrams involve 100 percent of your brain power. Picture the steps to solve difficult math problems in your mind.

Example: Use the FOIL method (see Figure X, on the facing page) to visually learn how to multiply binomials. Memorize the face until you can sketch it from memory. If you need to use it during a test, you can then sketch the face onto your scratch paper and refer to it.

Examples: When learning the *commutative property*, remember that the word *commutative* sounds like the word *community*. A community is made up of different types of people who could be labeled *as* an *a* group and a *b* group. However, in a community of *a* people and *b* people, it does not matter if we count the *a* people first or the *b* people first; we still have the same total number of people in the community. Thus, $a + b = b + a$.

When learning the *distributive law of multiplication over addition,* such as $a(b + c)$, remember that "distributive" sounds like "distributor," which is associated with giving out a product. The distributor *a* is giving its products to *b* and *c*.

Make Associations

Association learning can help you remember better. Find a link between new facts and some well-established old facts and study them together. The recalling of old facts will help you remember the new facts and strengthen a mental connection between the two. Make up your own associations to remember math properties and laws.

REMEMBER

The more ridiculous the association, the more likely you are to remember it.

Use Mnemonic Devices

The use of mnemonic devices is another way to help you remember. Mnemonic devices are easily remembered words, phrases, or rhymes associated with difficult-to-remember principles or facts.

Example: Many students become confused with the *order of operations*. These students mix up the order of the steps in solving a problem, such as dividing before adding the numbers

continued

in the parentheses. A mnemonic device to remember the order of operations is "Please Excuse My Dear Aunt Sally." The first letter in each of the words represents the math function to be completed from the first to the last. Thus, the order of operations is Parentheses (*Please*), Exponents (*Excuse*), Multiplication (*My*), Division (*Dear*), Addition (*Aunt*), and Subtraction (*Sally*). Remember to multiply and/or divide whatever comes first, from left to right. Also, add or subtract whatever comes first, from left to right.

Use Acronyms

Acronyms are another memory device to help you learn math. Acronyms are word forms created from the first letters of a series of words.

Example: FOIL is one of the most common math acronyms. FOIL is used to remember the procedure for multiplying two binomials. Each letter in the word *FOIL* represents a math operation. **FOIL** stands for **F**irst, **O**utside, **I**nside, and **L**ast, as it applies to multiplying two binomials such as $(2x + 3)(x + 7)$. The **F**irst quantities are $2x$ (in the first expression) and x (in the second expression). The **O**utside quantities are $2x$ (in the first

continued

expression) and 7 (in the second expression). The **I**nside quantities are 3 (in the first expression) and x (in the second expression). The **L**ast quantities are 3 (in the first expression) and 7 (in the second expression). This results in

$$\mathbf{F}\,(2x)(x) + \mathbf{O}\,(2x)(7) + \mathbf{I}\,(3)(x) + \mathbf{L}\,(3)(7).$$

Do the multiplication to get $2x^2 + 14x + 3x + 21$, which adds up to $2x^2 + 17x + 21$. See Figure 13 (The FOIL Method).

Section Review

1. How can a good math attitude help you learn?

_____

_____

_____

_____

2. Give an example of being a selective learner in your math class.

_____

_____

_____

_____

3. From your current math lessons, make up and explain one association remembering device that is not in this workbook.

_____

_____

_____

_____

4. For your next major math test, make up and explain one mnemonic device that is not in this workbook.

_____

_____

_____

_____

5. For your next major math test, make up and explain one acronym that is not in this workbook.

_____

_____

_____

_____

How to develop practice tests

Developing a practice test is one of the best ways to evaluate your memory and math skills before taking a real test. You want to find out what you do not know *before* the real test instead of *during* the test. Practice tests should be as real as possible and should include time constraints.

You can create a practice test by reworking all the problems since your last test that you have recorded in your problem log. Another practice test can be developed using every other problem in the textbook chapter tests. Further, you can use the solutions manual to generate other problems with which to test yourself. You can also use old exams from the previous semester. Check to see if the math lab/LRC or library has tests on file from previous semesters, or ask your instructor for other tests. For some students, a better way to prepare for a test is the group method.

Example: Hold a group study session several days before the test. Have each student prepare a test with ten questions. On the back of the test, have the answers listed, worked out step by step. Have each member of the study group exchange his or her test with another member of the group. Once all the tests have been completed, have the author of each test discuss with the group the procedures used to solve those problems.

If group work improves your learning, you may want to hold a group study session at least once a week. Make sure the individual or group test is completed at least three days before the real test.

Completing practice math tests will help you increase your testing skills. It will also reveal your test problem weaknesses in enough time for you to learn how to solve the problem before the real test. If you have difficulty with any of the problems during class or after taking the practice test, be sure to see your tutor or instructor.

After taking the practice test(s), you should know what parts you do not understand (and need to study) and what is likely to be on the test. Put this valuable information on one sheet of paper. This information needs to be understood and memorized. It may include formulas, rules, or steps to solving a problem.

Use the learning strategies discussed in this chapter to remember this information. A good example of how this information should look is what students might call a mental "cheat sheet." Obviously, you cannot use the written form of this sheet during the real test.

If you cannot take a practice test, put down on your mental cheat sheet the valuable information you will need for the test. Work to understand and memorize your mental cheat sheet. Chapter 7, "How to Improve Your Math Test-Taking Skills," will discuss how to use the information on the mental cheat sheet — *without cheating.*

Section Review

1. List three different ways you can make up a practice test.

First Way: _____

_____

_____

Second Way: _____

_____

_____

Third Way: _____

_____

_____

2. List ten different problems you would put on your practice test for the next exam.

First Problem: _____

_____

Second Problem: _____

_____

Third Problem: _____

_____

Fourth Problem: _____

_____

Fifth Problem: _____

_____

Sixth Problem: _____

_____

Seventh Problem: _____

_____

Eighth Problem: _____

_____

Ninth Problem: _____

_____

Tenth Problem: _____

_____

3. Give these ten problems to your study buddy and get ten problems from him or her. Compare your answers with those of your study buddy. How many did you get right? _____ How can you correct the wrong answers?

_____

_____

_____

_____

_____

_____

How to use
number sense

When taking your practice tests, you should use "number sense," or estimations, to make sure your answer is reasonable. Number sense is like common sense but it applies to math. Number sense is the ability to see if your answer makes sense without using algorithms. (Algorithms are the sequential math steps used to solve problems.) These following examples demonstrate solving a math problem (from a national math test given to high school students) using algorithms and number sense.

Example: Solve 3.04 x 5.3. Students use algorithms to solve this problem by multiplying the number 3.04 by 5.3, in sequence. Seventy-two percent of the students answered the problem

continued

correctly using algorithms.

Example: Estimate the product 3.04 x 5.3, and choose from the following answers.

A)   1.6   C) 160
B)   16   D) 1600

Only 15 percent of the students chose B, which is the correct answer. Twenty-eight percent of the students chose A. Using *estimating* to solve the answer, a whopping 85 percent of the students got the problem wrong.

These students were incorrectly using their "mental chalkboard" instead of using number sense. In using number sense to answer, you would multiply the numbers to the left of the decimal in each number to get an estimate of the answer. To estimate the answer, then, you would multiply 3 (the number to the left of the decimal in 3.04) by 5 (the number to the left of the decimal in 5.3) and expect the answer to be a little larger than 15.

It appears that the students' procedural processing (the use of algorithms) was good, but when asked to solve a non-routine problem using estimating (which is easier than using algorithms), the results were disappointing.

Another example of using number sense, or estimating, is in "rounding off."

Taking the time to estimate the answer to a math problem is a good way to check your answer. Another way to check your answer is to see if it is reasonable. Many students forget this important step and get the answer wrong. This is especially true of word or story problems.

Example: Solve 48 + 48 by rounding off. Rounding off means mentally changing the number (up or down) to make it more manageable to you, without using algorithms. By rounding off, 48 becomes 50 (easier to work with). 50 + 50 = 100. If the choices for answers were 104, 100, 98 and 96, you would then subtract 4 from the 100 (since each number was rounded up by 2) and you would get 96.

Also, when dealing with an equation, make sure to put the answer back into the equation to see if one side of the equation equals the other. If the two sides are not equal, you have the wrong answer. If you have extra time left over after you have completed a test, you should check answers using this method.

Examples: When solving a rate-and-distance problem, use your common sense to realize that one car cannot go 500 miles per hour to catch the other car. However, the car can go 50 miles per hour.

The same common sense rule applies to age-word problems, where the age of a person cannot be 150 years, but it can be 15 years.

Further, in solving equations, x is *usually* a number that is less than 20. When you solve a problem for x and get 50, then, this isn't reasonable, and you should recheck your calculations.

continued

Example: In solving the equation $x + 3 = 9$, you calculated that $x = 5$. To check your answer, substitute x with 5 and see if the problem works out correctly. 5 + 3 does not equal 9, so you know you have made a mistake and need to recalculate the problem. The correct answer, by the way, is $x = 6$.

REMEMBER

Number sense is a way to get more math problems correct by estimating your answer to determine if it is reasonable.

Section Review

1. Give an explanation of number sense.

_____

_____

_____

_____

_____

_____

_____

_____

2. How does number sense compare to common sense?

_____

_____

_____

_____

3. Give an example of a recent math problem you missed due to poor number sense.

_____

_____

_____

_____

4. List two ways you can improve your number sense.

First Way: _____

_____

_____

Second Way: _____

_____

_____

Name: _____ Date: _____

Chapter 6 Review

1. The way you receive information is through your five senses, which are _____, _____, _____, _____, and _____.

2. The main problem most students have is converting learned material into _____ _____ and _____ it.

3. Repeatedly reading and writing math material is one of the best ways for a _____ learner to study.

4. Reading _____ _____ is one of the best auditory ways to learn material.

5. A _____/_____ learner needs to feel and touch material to learn it.

6. Being a selective learner means _____ the material to study and learning the laws and principles _____ percent.

7. Mnemonic devices are easy-to-remember _____, _____, or _____ associated with difficult-to-remember principles or facts.

8. _____ are word forms created from the first letters of a series of words.

9. The reason to develop a practice test is to find out what you do not know _____ the test instead of _____ the test.

10. Social learners benefit from forming _____ _____.

What is the most important information you learned from this chapter?

_____

_____

_____

_____

_____

_____

_____

How can you immediately use it?

_____

_____

_____

_____

_____

_____

_____

How to Improve Your Math Test-Taking Skills

Taking a math test is different from taking tests in other subjects. Math tests require not only that you recall the information, but that you apply the information. Multiple-choice tests, for example, usually test you on recall, and if you do not know the answer, you can guess.

Math tests build on each other, whereas history tests often do not test you on previous material. Most math tests are speed tests, where the faster your are, the better grade you can receive; most social science tests are designed for everyone to finish.

Math test preparation and test-taking skills are different from preparation and skills needed for other tests. You need to have a test-taking plan and a test analysis plan to demonstrate your total knowledge on math tests. Students with these plans make better grades compared to students without them. Math instructors want to measure your math knowledge, not your poor test-taking skills.

In this chapter, you will learn

- why attending class and doing your homework may not be enough to pass,
- the general pretest rules,
- the ten steps to better test taking,
- the six types of test-taking errors, and
- how to prepare for the final exam.

Why attending class and doing your homework may not be enough to pass

Most students and some instructors believe that attending class and doing all the homework ensures an A or B on tests. This is far from true. Doing all the homework and getting the correct answers is very different in many ways from taking tests:

1. While doing homework, there is little anxiety. A test situation is just the opposite.

2. You are not under a time constraint while doing your homework; you may have to complete a test in 55 minutes or less.

3. If you get stuck on a homework problem, your textbook and notes are there to assist you. This is not true for most tests.

4. Once you learn how to do several problems in a homework assignment, the rest are similar. In a test, the problems may be in random order.

5. In doing homework, you have the answers to at least half the problems in the back of the text and answers to all the problems in the solutions guide. This is not true for tests.

6. While doing homework, you have time to figure out how to correctly use your calculator. During the test, you can waste valuable time figuring out how to use your calculator.

7. When doing homework, you can call your study buddy or ask the tutor for help, something you cannot do on the test.

8. While doing your homework, you can go to websites for online tutoring to help you solve problems you do not understand.

Do not develop a false sense of security by believing you can make an A or B by doing just your homework. Tests measure more than just your math knowledge.

The general pretest rules

General rules are important when taking any type of test:

1. *Get a good night's sleep before taking a test.* This is true for the ACT, the SAT, and your math tests. If you imagine you are going to cram all night and perform well on your test with three to four hours of sleep, you are wrong. It would be better to get seven or eight hours of sleep and be fresh enough to use your memory to recall information needed to answer the questions.

2. *Start studying for the test at least three days ahead of time.* Make sure you take a practice test to find out, before the test, what you do not know. Review your problem log and work the problems. Review the concept errors you made on the last test. (How to identify and correct your concept errors will be discussed later in this chapter.) Meet with the instructor or tutor for help on those questions you cannot solve.

3. *Review only already-learned material the night before a test.*

4. *Make sure you know all the information on your mental cheat sheet.* Review your notebook and glossary to make sure you understand the concepts. Work a few problems and recall the information on your mental cheat sheet right before you go to bed. Go directly to bed; do not watch television, listen to the radio, or party. While you are asleep, your mind will work on and remember the last thing you did before going to sleep.

5. *Get up in the morning at your usual time and review your notes and problem log.* Do not do any new problems, but make sure your calculator is working.

Section Review

1. List three reasons why only attending class and doing your homework may not be enough to pass your math course.

First Reason: _____

_____

_____

Second Reason: _____

_____

_____

Third Reason: _____

_____

_____

2. List and explain three general pretest rules that best apply to you.

First Rule: _____

_____

_____

Second Rule: _____

_____

_____

Third Rule: _____

_____

_____

The ten steps to better test taking

You need to have a game plan to take a math test. This plan is different from taking history, English, humanities and some science tests. The game plan is to get the most points in the least amount of time. Many students lose test points because they use the wrong test-taking strategies for math. By following these ten steps you can demonstrate more knowledge on the test and get more problems right.

Step 1 *Use a memory data dump.* When you get your test, turn it over and write down the information that you might forget. Remember, this is your mental cheat sheet that you should have already made while preparing for the test. Your mental cheat sheet has now turned into a mental list, and writing down this information is not cheating. Do not put your name on the test, do not skim it, just turn it over and write down those facts, figures, and formulas from your mental cheat sheet or other information you might not remember during the test. This is called your *first memory data dump.* The data dump provides memory cues for test questions.

Example: It might take you a while to remember how to do a coin problem. However, if you had immediately turned your test over and written down different ways of solving coin problems, it would be easier to solve the coin problem.

Step 2 *Preview the test.* Put your name on the test and start previewing. Previewing the test requires you to look through the entire test to find different types of problems and their point values. Put a mark by the questions that you can do without thinking. These are the questions that you will solve first.

Step 3 *Do a second memory data dump.* The secondary data dump is for writing down material that was jarred from your memory while previewing the test. Write this information on the back of the test.

Step 4   *Develop a test progress schedule.* When you begin setting up a test schedule, determine the point value for each question. Some test questions might be worth more points than others.

In some tests, word problems are worth 5 points and other questions might be worth 2 or 3 points. You must decide the best way to get the most points in the least amount of time. This might mean working the questions worth 2 or 3 points first and leaving the more difficult word problems for last.

Decide how many problems should be completed halfway though the test. You should have more than half the problems completed by that time.

Step 5   *Answer the easiest problems first.* Solve, in order, the problems you marked while previewing the test. Then review the answers to see if they make sense. Start working through the test as fast as you can while being accurate. Answers should be reasonable.

Example: The answer to a problem asking you to find the area of a rectangle cannot be negative, and the answer to a land-rate-distance problem cannot be 1000 miles per hour.

Clearly write down each step to get partial credit. Even if you end up missing the problem, you might get some credit. In most math tests, the easier problems are near the beginning of the first page; you need to answer them efficiently and quickly. This will give you both more time for the harder problems and time to review.

Step 6   *Skip difficult problems.* If you find a problem that you do not know how to work, read it twice and automatically skip it. Reading it twice will help you understand the problem and put it into your working memory. While you are solving other problems, your mind is still working on that problem. Difficult problems could be the type of problem you have never seen before or a problem in which you get stuck on the second or third step. In either case, skip the problem and go on to the next one.

Step 7   *Review the skipped problems.* When working the skipped problems, think how you have solved other, similar problems as a cue to solving the skipped ones. Also try to remember how the instructor solved that type of problem on the board.

While reviewing skipped problems, or at any other time, you may have the "Aha!" response. The "Aha!" response is your remembering how to do a skipped problem. Do not wait to finish your current problem. Go to the problem on which you had the "Aha!" and finish that problem. If you wait to finish your current problem, your "Aha!" response could turn into an "Oh, no!" response.

Step 8   *Guess at the remaining problems.* Do as much work as you can on each problem, even if it is just writing down the first step. If you cannot write down the first step, rewrite the problem. Sometimes rewriting the problem can jar your memory enough to do the first step or the entire problem. This is true of students, particularly of tactile/concrete learners where writing can trigger the memory process of solving the problem. Remember, you did not learn how to solve math problems with your hands tied behind your back. If you leave the problem blank, you will get a zero. Do not waste too much time on guessing or trying to work the problems you cannot do.

Step 9   *Review the test.* Look for careless errors or other errors you may have made. Students usually lose 2 to 5 test points on errors that could have been caught in review. Do not talk yourself out of an answer just because it may not look right. This often happens when an answer does not come out even. It is possible for the answer to be a fraction or decimal.

REMEMBER

Answers in math do not have "dress codes." Research reveals that the odds of changing a right answer to a wrong answer are greater than the odds of changing a wrong answer to a right one.

Step 10   *Use all the allowed test time.* Review each problem by substituting the answer back into the equation or doing the opposite function required to answer the question. If you cannot check the problem by the two ways mentioned, rework the problem on a separate sheet of paper and compare the answers. Do not leave the test room unless you have reviewed each problem two times or until the bell rings.

Even though we encourage students to work until the end of the test period most students leave the classroom before the end of the period. These students state that even though they know they should use all the test time, they cannot stay in the room until the end of the test time. These students also know that their grades would probably improve if they kept checking their answers or kept working the problems they are having difficulty with. After talking to hundreds of these students, I discovered two different themes for leaving the classroom early. First, test anxiety gets so overwhelming that they cannot stay in the room. The relief from the test anxiety (leaving the room) is worth more than getting a better grade. If you are one of these students, you must learn how to decrease your test anxiety by following the suggestions in Chapter 4 or by using the *How to Reduce Test Anxiety* cassette tape. Don't let the anxiety control your grades!

The other reason for leaving the test early is that they do not want to be the last or one of the last few students to turn in their tests. They still believe that students who turn their tests in last are "dumb and stupid." These students also believe that students who turn their tests in first make "A's" and "B's" and those students who turn their tests in last make "D's" and "F's." If you are one of these students, you don't need to care about what other students think about you (it's usually wrong anyway). YOU need to fight the urge to leave early and use all the test time. Remember, passing mathematics is the best way to get a high paying job and support yourself or your family. DO IT NOW!

REMEMBER

There is no prize for handing your test in first, and students who turn their papers in last do make A's.

Stapling your scratch paper to the math test when handing it in has several advantages:

- If you miscopied the answer from the scratch paper, you will probably get credit for the answer.
- If you get the answer incorrect due to a careless error, your work on the scratch paper could give you a few points.
- If you do get the problem wrong, it will be easier to locate the errors when the instructor reviews the test. This will prevent you from making the same mistakes on the next math test.

REMEMBER

Handing in your scratch paper may get you extra points or improve your next test score.

Section Review

1. List and explain the ten steps to test taking.

   First Step: _____

   _____

   Second Step: _____

   _____

   _____

Third Step: _____

_____

Fourth Step: _____

_____

Fifth Step: _____

_____

Sixth Step: _____

_____

Seventh Step: _____

_____

Eighth Step: _____

_____

Ninth Step: _____

_____

Tenth Step: _____

_____

2. After trying the ten steps to test taking, develop your own personalized test-taking steps.

   First Step: _____

   _____

   Second Step: _____

   _____

   Third Step: _____

   _____

   Fourth Step: _____

   _____

Fifth Step: _____

_____

Sixth Step: _____

_____

Seventh Step: _____

_____

Eighth Step: _____

_____

Ninth Step: _____

_____

Tenth Step: _____

_____

Additional steps

_____

_____

_____

_____

_____

The six types of test-taking errors

To improve future test scores, you must conduct a test analysis of previous tests. In analyzing your tests, you should look for the following kinds of errors:

1. misread-directions errors
2. careless errors
3. concept errors
4. application errors
5. test-taking errors
6. study errors

Students who conduct math test analyses will improve their total test scores.

Misread-directions errors occur when you skip directions or misunderstand directions, but do the problem anyway.

Examples: Suppose you have this type of problem to solve:

$$(x + 1)(x + 1)$$

Some students will try to solve for x, but the problem only calls for multiplication. You would solve for x only if you have an equation such as $(x + 1)(x + 1) = 0$.

Another common mistake is not reading the directions before doing several word problems or statistical problems. All too often, when a test is returned, you find only three out of the five problems had to be completed. Even if you did get all five of them correct, it cost you valuable time that could have been used obtaining additional test points.

To avoid misread-direction errors, carefully read and interpret all the directions. Look for anything that is unusual, or if the directions have two parts. If you do not understand the directions, ask the instructor for clarification. If you feel uneasy about asking the instructor for interpretation of the question, remember the instructor in most cases does not want to test you on the interpretation of the question but how you answer it. Also, you don't want to make the mistake of assuming that the instructor will not interpret the question. Let the instructor make the decision to interpret the question, not you.

Careless errors are mistakes that you can catch automatically upon reviewing the test. Both good and poor math students make careless errors. Such errors can cost a student a higher letter grade on a test.

Examples: *Dropping the sign:*
$-3(2x) = 6x$, instead of $-6x$
which is the correct answer.
Not simplifying your answer:
Leaving $(3x - 12)/3$ as your answer instead of simplifying it to $x - 4$.
Adding fractions: $1/2 + 1/3 = 2/5$, instead of $5/6$, which is the correct answer.
Word problems: $x = 15$ instead of "The student had 15 tickets."

However, many students want all their errors to be careless errors. This means that the students *did* know the math, but simply made silly mistakes. In such cases, I ask the student to solve the problem immediately, while I watch.

If the student can solve the problem or point out his or her mistake in a few seconds, it is a careless error. If the student cannot solve the problem immediately, it is not a careless error and is probably a concept error.

When working with students who make careless errors, I ask them two questions: First,

"How many points did you lose due to careless errors?" Then I follow with, "How much time was left in the class period when you handed in your test?" Students who lose test points to careless errors are giving away points if they hand in their test papers before the test period ends.

To reduce careless errors, you must realize the types of careless errors made and recognize them when reviewing your test. If you cannot solve the missed problem immediately, it is not a careless error. If your major error is not simplifying the answer, review each answer as if it were a new problem and try to reduce it.

Concept errors are mistakes made when you do not understand the properties or principles required to work the problem. Concept errors, if not corrected, will follow you from test to test, causing a loss of test points.

Examples: Some common concept errors are not knowing:
$(-)(-)x = x, \text{ not } -x$
$-1(2) > x(-1) = 2 < x, \text{ not } 2 > x$
$5/0 \text{ is undefined, } not \, 0$
$(a + x) / x \text{ is } not \text{ reduced to } a$
the order of operations

Concept errors must be corrected to improve your next math test score. Students who have numerous concept test errors will fail the next test and the course if concepts are not understood. Just going back to rework the concept error problems is not good enough. You must go back to your textbook or notes and learn why you missed those types of problems, not just the one problem itself.

The best way to learn how to work those types of problems is to set up a concept problem error page in the back of your notebook. Label the first page "Test One Concept Errors." Write down all your concept errors and how to solve the problems. Then, work five more problems that use the same concept. Now, in your own words, write the concepts that you are using to solve these problems.

If you cannot write the concept in your own words, you do not understand it. Get assistance from your instructor if you need help finding similar problems using the same concept or cannot understand the concept. Do this for every test.

Application errors occur when you know the concept but cannot apply it to the problem. Application errors usually are found in word problems, deducing formulas (such as the quadratic equation), and graphing. Even some better students become frustrated with application errors; they understand the material but cannot apply it to the problem.

To reduce application errors, you must predict the type of application problems that will be on the test. You must then think through and practice solving those types of problems using the concepts.

> **Example:** If you must derive the quadratic formula, you should practice doing that backward and forward while telling yourself the concept used to move from one step to the next.

Application errors are common with word problems. After completing the word problem, reread the question to make sure you have applied the answer to the intended question. Application errors can be avoided with appropriate practice and insight.

Test-taking errors apply to the specific way you take tests. Some students consistently make the same types of test-taking errors. Through recognition, these bad test-taking habits can be replaced by good test-taking habits. The result will be higher test scores. The list that follows includes the test-taking errors that can cause you to lose many points on an exam:

1. *Missing more questions in the first third, second third, or last third of a test* is considered a test-taking error.

   Missing more questions in the first third of a test can be due to carelessness when doing easy problems or due to test anxiety.

   Missing questions in the last part of the test can be due to the fact that the last problems are more difficult than the earlier questions or due to increasing your test speed to finish the test.

   If you consistently miss more questions in a certain part of the test, use your remaining test time to review that section of the test first. This means you may review the last part of your test first.

2. *Not completing a problem to its last step* is another test-taking error. If you have this bad habit, review the last step of the test problem first, before doing an in-depth test review.

3. *Changing test answers from correct ones to incorrect ones* is a problem for some students. Find out if you are a good or bad answer-changer by comparing the number of answers changed to correct and to incorrect answers. If you are a bad answer-changer, write on your test, "Don't change answers." Change answers only if you can prove to yourself or the instructor that the changed answer is correct.

4. *Getting stuck on one problem and spending too much time on it* is another test-taking error. You need to set a time limit on each

problem before moving to the next problem. Working too long on a problem without success will increase your test anxiety and waste valuable time that could be used in solving other problems or in reviewing your test.

5. *Rushing through the easiest part of the test and making careless errors* is a common test-taking error for the better student. If you have the bad habit of getting more points taken off for the easy problems than for the hard problems, first review the easy problems, and later review the hard problems.

6. *Miscopying an answer from your scratch work to the test* is an uncommon test-taking error, but it does cost some students points. To avoid these kinds of errors, systematically compare your last problem step on scratch paper with the answer written on the test. In addition, always hand in your scratch work with your test.

7. *Leaving answers blank* will get you zero points. If you look at a problem and cannot figure out how to solve it, do not leave it blank. Write down some information about the problem, rewrite the problem, or try to do at least the first step.

REMEMBER

Writing down the first step of a problem is the key to solving the problem and obtaining partial credit.

8. *Solving only the first step of a two-step problem* causes problems for some students. These students get so excited when answering the first step of the problem that they forget about the second part. This is especially true on two-step word problems. To correct this test-taking error, write "two" in the margin of the problem. That will remind you that there are two steps or two answers to this problem.

9. *Not understanding all the functions of your calculator* can cause major testing problems. Some students only barely learn how to use the critical calculator functions. They then forget or have to relearn how to use their calculator, which costs test points and test time. Do not wait to learn how to use your calculator on the test. Overlearn the use of your calculator *before* the test.

10. *Leaving the test early without checking all your answers is a costly habit.* Do not worry about the first person who finishes the test and leaves. Many students start to get nervous when students start to leave after finishing the test. This can lead to test anxiety, mental blocks, and loss of recall.

According to research, the first students finishing the test do not always get the best grades. It sometimes is the exact opposite. Ignore the exiting students, and always use the full time allowed.

Make sure you follow the ten steps to better test taking. Review your test taking procedures for discrepancies in following the ten steps to better test taking. Deviating from these proven ten steps will cost you points.

Study errors, the last type of mistake to look for in test analysis, occur when you study the wrong type of material or do not spend enough time on pertinent material. Review your test to find out if you missed problems because you did not practice that type of problem or because you did practice it but forgot how to do it during the test. Study errors

Figure 17

Math Test for Prealgebra

The answers are listed in bold. The correct answer on missed questions is shaded. Identify the type of error based on the "6 Types of Test-taking Errors." The student's test score is 70.

1. Write in words: 32.685

   Thirty-two and six hundred eighty-five thousandths

2. Write as a fraction and simplify: 0.078

   78   **39**
   1000   **500**

3. Round to the nearest hundredth: 64.8653

   64.865   **64.87**   (-4)

4. Combine like terms:

   6.78x -3.21 + 7.23x -6.19

   = 6.78x + 7.23x + (-3.21) + (-6.19)
   = 14.01x -9.4

5. Divide and round to the nearest hundredth: 68.1357   2.1

   32.445 → 32.45

6. Write as a decimal: $\frac{5}{16}$

   0.3125

7. Insert < or > to make a true statement:
   $\frac{3}{8}$ < $\frac{6}{13}$

8. Solve: $\frac{3}{x} = \frac{9}{12}$

   9x = 3(12)

   $\frac{9x}{9} = \frac{36}{9}$

   x = 5 $x = 4$

9. What number is 35% of 60?

   2100   **21.00**   (-4)

10. 20.8 is 40% of what number?   52

    (-8)

11. 567 is what percent of 756?

    $\frac{756}{567} = \frac{9}{100}$

    = 133.3%   75%   (-8)

12. Multiply: (-2)

    (-6.03) (-2.31) = 13.9   **13.9293**

continued

Answer Key for Exercise

1. Correct
2. *Misread directions* — Forgot to simplify by reducing the fraction.
3. *Concept error* — Did not know that hundredths is two places to the right of the decimal.
4. Correct
5. Correct
6. Correct
7. Correct
8. *Careless error* — Divided incorrectly in the last step.
9. *Test-taking error* — Not following step 5 in test-taking steps: Reviewing answers to see if they make sense. The number that equals 35% can't be larger than 60.
10. *Test-taking error* — Did not follow step 7 in test taking steps: Don't leave an answer blank
11. *Application error* — Solved the equation correctly but did not apply the correct equation.
12. *Concept error* — Did not know that when you multiply with one number in the hundredths, the answer must include the hundredths column.

will take some time to track down. But correcting study errors will help you on future tests.

Most students, after analyzing one or several tests, will recognize at least one major, common test-taking error. Understanding the effects of this test-taking error should change your study techniques or test-taking strategy.

Example: If there are seven minutes left in the test, should you review for careless errors or try to answer those two problems you could not totally solve? This is a trick question. The real question is, Do you miss more points due to careless errors or concept errors, or are the missed points about even? The answer to this question should determine how you will spend the last minutes of the test. If you miss more points due to careless errors or miss about the same number of points due to careless/concept errors, review for careless errors.

Careless errors are easier to correct than concept errors. However, if you make very few or no careless errors, you should be working on those last two problems to get the greatest number of test points. Knowing your test-taking errors can add more points to your test by changing your test-taking procedure.

Section Review

1. List the six types of test taking errors.

   Error One: _____

   Error Two: _____

   Error Three: _____

   Error Four: _____

   Error Five: _____

   Error Six: _____

2. Which of these errors have you made in the past?

   _____

   _____

   _____

   _____

   _____

   _____

3. How can you avoid these types of errors on the next test?

   _____

   _____

   _____

   _____

   _____

   _____

4. Review the Math Test for Prealgebra if you are in a prealgebra course. Review for type of errors, the points lost for each error, and example of each error.

   Error Type: _____

   Points Lost for Error: _____

   Example of Error: _____

   _____

   Error Type: _____

   Points Lost for Error: _____

   Example of Error: _____

   _____

   Error Type: _____

   Points Lost for Error: _____

   Example of Error: _____

   _____

   Error Type: _____

   Points Lost for Error: _____

   Example of Error: _____

   _____

   Error Type: _____

   Points Lost for Error: _____

   Example of Error: _____

   _____

   Error Type: _____

   Points Lost for Error: _____

   Example of Error: _____

   _____

Error Type: _____

Points Lost for Error: _____

Example of Error: _____

_____

5. Review your last test to list the type of errors, the points lost for each error, and examples of the errors. You may have more errors than allowed for in the list below. Use a separate sheet of paper if you need to list more. (Remember: Careless errors are problems you can solve immediately.)

Error Type: _____

Points Lost for Error: _____

Example of Error: _____

_____

Error Type: _____

Points Lost for Error: _____

Example of Error: _____

_____

Error Type: _____

Points Lost for Error: _____

Example of Error: _____

_____

Error Type: _____

Points Lost for Error: _____

Example of Error: _____

_____

6. List any additional concept error problems that were not included in question 2.

Points Lost for Error: _____

Example of Error: _____

_____

Points Lost for Error: _____

Example of Error: _____

_____

Using the pages in the back of your math notebook, rework the concept error problems and do five math problems just like them. For each concept error, write why you could not solve the problem on the test and what new information you learned to solve the problem. Do this for every major test. Give an example of one of your concept error(s) and the reasons you can now solve it.

Concept Error Problem(s):

_____

_____

_____

_____

_____

Reasons I Can Now Solve It:

_____

_____

_____

_____

_____

How to prepare for the final exam

The first day of class is when you start preparing for the final exam. Look at the syllabus or ask the instructor if the final exam is cumulative. A cumulative exam covers everything from the first chapter to the last chapter. Most math final exams are cumulative.

The second question you should ask is if the final exam is a departmental exam or if it is made up by your instructor. In most cases, departmental exams are more difficult and need a little different preparation. If you have a departmental final, you need to ask for last year's test and ask other students what their instructors say will be on the test.

The third question is, How much will the final exam count? Does it carry the same weight as a regular test or, as in some cases, will it count a third of your grade? If the latter is true, the final exam will usually make a letter grade difference on your final grade. The final exam could also determine if you pass or fail the course. Knowing this information before the final exam will help you prepare.

Preparing for the final exam is similar to preparing for each chapter test. You must create a pretest to discover what you have forgotten. You can use questions from the textbook chapter tests or questions from your study group.

Review the concept errors that you recorded in the back of your notebook labeled "Test One," "Test Two," and so on. Review your problem log for questions you consistently miss. Even review material that you knew for the first and second test but which was not used on any other tests. Students forget how to work some of these problems.

If you do not have a concept error page for each chapter and did not keep a problem log you need to develop a pre-test before taking the final exam. If you are individually preparing for the final, then copy each chapter test and do every fourth problem to see what errors you may make. However, it is better to have a study group where each of the four members brings in ten problems with the answers worked out on a separate page. Then each group member can take the 30-question test to find out what they need to study. You can refer to the answers to help you solve the problems that you miss. Remember you want to find out what you don't know before the test not during the test. Developing and using these methods before each test will improve your grades.

Make sure to use the 10 steps to better test taking and the information gained from your test analysis. Use all the time on the final exam because you could improve your final grade by a full letter if you make an "A" on the final.

Section Review

1. What are the three qustions you need to ask about your final exam?

First Question: _____

_____

_____

Second Question: _____

_____

_____

Third Question: _____

_____

_____

Chapter 7 Review

1. You should start studying for a test at least _____ days ahead of time.

2. You should be reviewing only already-_____ material the night before the test.

3. When previewing the test, put a _____ by the questions you can solve without thinking, and do those problems _____.

4. When taking a test, you must decide the best way to get the _____ points in the _____ amount of time.

5. Halfway through the test, you should have _____ than half of the problems completed.

6. If you cannot write down the first step of a problem, you should _____ the problem to help you remember how to solve it.

7. Do not leave the test room until you have reviewed each problem _____ times or until the bell rings.

8. In reviewing the test, if you cannot solve the problem immediately (within a few seconds), it is not a _____ error and is probably a _____ error.

9. Students who have numerous _____ errors will fail the next test and the course if the _____ are not understood.

10. The _____ day of class is when you start preparing for the final exam.

What is the most important information you learned from this chapter?

_____
_____
_____
_____
_____
_____
_____
_____

How can you immediately use it?

_____
_____
_____
_____
_____
_____
_____
_____

Appendix: Math Autobiography

This is the purpose for your math autobiography—to reflect on your past experiences when learning math, how they made you feel then, and how these experiences and feelings shape how you feel about what you are doing in your current math classes. It will close with solutions you believe will help you improve your experiences in math classes.

  The following writing process will simplify this task while improving the final paper.

1. Use the questions in the sections on types of math anxiety and causes of math anxiety in Chapter 3 to brainstorm ideas for your paper. Record these answers on paper.

2. By each experience you record, write how the experience made you feel.

3. To further your brainstorming, complete the following sentences:

My favorite math instructor was

_____,

because he/she

_____

_____

_____,

I always felt

_____

_____

_____.

One math instructor who bothered me was

_____,

because he/she

_____

_____

_____,

I always felt

_____

_____

_____.

continued

When I hear the word math, I think of

_____

_____

_____

_____

I did well in math until

_____

_____

_____

_____

The first time I experienced math anxiety was when

_____

_____

_____

_____

The first time I was anxious before or during a math test was _____

_____

_____

_____

_____

The most recent time I was anxious about math class was when

_____

_____

_____

_____

_____

_____

The most recent time I was anxious before or during a test was when

_____

_____

_____

_____

_____

4. Since an autobiography is in chronological order, identify the important information in your brainstorming and place into the correct time category in the tables on the following pages. These tables will be the outline for your paper.

5. You will write the introduction last, after you have thought through the meaning of your experiences. Then, when you type or write the final paper, you will place the introduction paragraph first.

Elementary School
Main Paragraph One

Experiences	Feelings

As a result of these experiences, now when I am either in math class, stydying for math, or taking a math test, I . . .

Middle School
Main Paragraph Two

Experiences	Feelings

As a result of these experiences, now when I am either in math class, studying for math, or taking a math test, I . . .

High School
Main Paragraph Three

Experiences	Feelings

As a result of these experiences, now when I am either in math class, studying for math, or taking a math test, I . . .

College and/or Work Experience
Main Paragraph Four (If this applies)

Experiences	Feelings

As a result of these experiences, now when I am either in math class, studying for math, or taking a math test, I . . .

Conclusion Paragraph

When I look back on my experiences learning or using math, I understand why I have developed attitudes like

_____

_____

_____ .

To develop a more positive perspective toward math, I need to improve certain behaviors.

Behaviors	*Solutions for Change*
1.	
2.	
3.	
4.	
As a result, I will become . . .	

Introduction Paragraph

(When you write your final paper, this paragraph will be first. This paragraph should introduce the main lesson you have learned through this activity. Look at what you have written for your conclusion paragraph for some ideas.)

6. Now, using the tables as your outline, either handwrite (if allowed by the instructor) or type the final paper to be turned in to your instructor.

Bibliography

Bloom, B. 1976. *Human Characteristics and School Learning.* New York: McGraw-Hill Book Company.

Brown, J. F., and R. M. Cooper. 1978. *Learning Styles Inventory.* Freeport, NY: Educational Activities, Inc.

Cattell, R. B. 1 966. "The Screen Test for the Number of Factors." *Multivariate Behavioral Research,* no. 1, pp. 245-276.

Fox, L. H. 1977. *Women and Mathematics: Research Perspective for Change.* Washington, DC: National Institute of Education.

Kolb, D. 1985. *Learning-Styles Inventory.* Boston, MA: McBer and Company.

McCarthy, B. 1981. *The 4-Mat System: Teaching to Learning Styles with Right/Left Mode Techniques.* Barrington, IL: Excel.

Nolting, P. D. 1987. *How to Reduce Test Anxiety.* An audiocassette tape. Bradenton, FL: Academic Success Press. (www.academicsuccess.com)

Richardson. F. C., and R. M. Suinn. 1973. "A comparison of Traditional Systematic Desensitization, Accelerated Mass Desensitization, and Mathematics Anxiety." *Behavior Therapy,* no. 4, pp. 212-218.

Suinn, R. M. 1988. *Mathematics Anxiety Rating Scale (MARS).* Fort Collins, CO: RMBSI, Inc.

Tobias, S. 1978. "Who's Afraid of Math and Why?" *Atlantic Monthly,* September, pp. 63-65.

Wolpe, J. 1958. *Psychotherapy by Reciprocal Inhibition.* Stanford, CA: Stanford University Press.

About the Author

Over the past 15 years, Learning Specialist Dr. Paul Nolting has helped thousands of students improve their math learning and obtain better grades. Dr. Nolting is a national expert in assessing math learning problems — from study skills to learning disabilities — and developing effective learning strategies and testing accommodations.

Dr. Nolting is also a nationally recognized consultant and trainer of math study skills and of learning and testing accommodations for students with learning disabilities. He has conducted national training grant workshops on math learning for the Association on Higher Education and Disabilities and for a two-year University of Wyoming 1991 Training Grant of Basic Skills.

Dr. Nolting has conducted numerous national conference workshops on math learning for the National Developmental Education Association, the National Council of Educational Opportunity Association, and the American Mathematical Association of Two-Year Colleges. He is a consultant for the American College Test (ACT) and the Texas Higher Education Coordinating Board. He is also a consultant for Houghton Mifflin faculty development programs, conducting workshops on math study skills for math faculty. His text *Winning at Math: Your Guide to Learning Mathematics Through Successful Study Skills* is used throughout the United States, Canada, and the world as the definitive text for math study skills.

Dr. Nolting has consulted with numerous universities and colleges. Some of the universities with which Dr. Nolting has consulted are the University of Massachusetts, the University of Colorado-Boulder, Texas Tech. University, Black Hills State University, Tennessee Tech. University, and the University of Connecticut.

Some of the colleges with which he has consulted are San Antonio College, St. Louis Community College, J. Sargeant Reynolds College, Montgomery College, Broward Community College, Miami-Dade Community College, Northeast State Technical Community College, Landmark College, Denver Community College, Valencia Community College, and Pensacola Junior College. Dr. Nolting has consulted with over 75 colleges, universities, and high schools over the last 15 years.

Dr. Nolting holds a Ph.D. in education from the University of South Florida. His Ph.D. dissertation was "The Effects of Counseling and Study Skills Training on Mathematics Academic Achievement." He is an adjunct instructor for the University of South

Florida, teaching assessment and appraisal courses at the graduate level.

His book *Winning at Math: Your Guide to Learning Mathematics Through Successful Study Skills* was selected Book of the Year by the National Association of Independent Publishers. "The strength of the book is the way the writer leads a reluctant student through a course from choosing a teacher to preparing for the final examination," says *Mathematics Teacher*, a publication of the National Council of Teachers of Mathematics. *Winning at Math* and Dr. Nolting's other titles are available at http://www.academicsuccess.com.

His two audiocassettes, *How to Reduce Test Anxiety* and *How to Ace Tests*, were also winners of awards in the National Association of Independent Publishers competition. "Dr. Nolting," says *Publisher's Report*, "is an innovative and outstanding educator and learning specialist."

A key speaker at numerous regional and national education conferences and conventions, Dr. Nolting has been widely acclaimed for his ability to communicate with faculty and students on the subject of improving math learning.

Books and Materials by Academic Success Press and Dr. Paul Nolting

How to Ace Tests. Academic Success Press, 1987.

How to Reduce Test Anxiety. Academic Success Press, 1987. (audiocassette)

How to Develop Your Own Math Study Skills Workshop and Course. Academic Success Press, 1991.

Improving Mathematics Studying and Test-Taking Skills. Boston, MA: Houghton Mifflin Company, 1994. (videotape, ISBN 0-669-33291-7)

Math and the Learning Disabled Student: A Practical Guide for Accommodations. Academic Success Press, 1991.

Math and Students with Learning Disabilities: A Practical Guide to Course Substitutions. Academic Success Press, 1993.

Strategy Cards for Higher Grades. Academic Success Press, 1989. Twenty 3-inch by 5-inch cards. (card deck)

Winning at Math — Computer Evaluation Software. Academic Success Press.

Winning at Math: Your Guide to Learning Mathematics Through Successful Study Skills. Academic Success Press, 1997.

Ordering Information

For ordering information on *Winning at Math* and other books and learning materials listed above, call toll free: 1-800-247-6553 or visit www.academicsuccess.com. Credit card, checks, and purchase orders are accepted. Shipping charges are $6.00 for the first item and $2.00 for any additional items. Academic Success Press can be contacted at 6023 26th Street West, PMB 132, Bradenton, FL 34207. For a free catalog, call (941) 359-2819.

To order Houghton Mifflin videos, contact customer service toll free at 1-800-225-1464.